The Fragile Contract

The Fragile Contract

University Science and the Federal Government

edited by David H. Guston and Kenneth Keniston

The MIT Press, Cambridge, Massachusetts, and London, England

This book was set in New Baskerville at The MIT Press and was printed and bound in the United States of America.

Library of Congress Cataloging-in-Publication Data

The Fragile contract / edited by David H. Guston and Kenneth Keniston.
p. cm.
Includes bibliographical references and index.
ISBN 0-262-07161-4. – ISBN 0-262-57107-2 (pbk.)
1. Research—Government policy—United States. 2. Science and state—United States. 3. Universities and colleges—Research—United States. 4. Research institutes—United States. I. Guston, David H. II. Keniston, Kenneth.
Q180.U5F73 1994
338.97306—dc20 94-20567
CIP

Contents

Acknowledgments

This volume grew out of a faculty workshop at the Massachusetts Institute of Technology (MIT) on the relationship between scientific research universities and the federal government. The workshop brought a wide range of speakers together with a group of senior MIT faculty members and administrators, many of whom were charged with negotiating, maintaining, and improving just the sorts of relationships between science and politics that were being addressed.

The format of the workshop was intended to encourage extensive interaction. Each speaker's talk was distributed to the participants in advance of the meetings, held monthly during academic year 1991–1992, often accompanied by additional readings. The workshops started at 5 p.m. Before dinner, the speaker summarized his or her paper, and discussion focused on questions of fact. After dinner, one of the regular workshop participants reopened discussion with prepared comments. The ensuing two hours were more like an extended conversation than a formal question-and-answer period, with lively and sometimes heated exchanges with the speaker and among workshop members. A detailed summary of the discussion was used by the speakers to help prepare the final versions of their papers for this volume. (The papers by Gerald Holton and by Peter Likins and Albert H. Teich were commissioned after the workshop.)

The editors have incurred many debts during the preparation of this volume. We gratefully acknowledge the support of the Carnegie

Corporation of New York and the Office of the Provost at MIT for providing funding for the workshops. Professor Sheila Widnall, then associate provost at MIT and presently Secretary of the Air Force, was instrumental in conceiving and encouraging the project.

MIT participants in the workshop included: Robert Birgenau (Dean of the School of Science and Professor of Physics), John C. Crowley (Special Assistant to the MIT President and Director of the MIT Washington Office), Richard de Neufville (Director of the Technology and Policy Program and Professor of Civil Engineering), Bernard J. Frieden (Ford Professor of Urban Studies and Planning), Jerome I. Friedman (Institute Professor, Physics), Morris Halle (Institute Professor, Philosophy and Linguistics), Nancy H. Hopkins (Professor of Biology), Henry D. Jacoby (William F. Pounds Professor of Management), Thomas H. Jordan (Head, Department of Earth Atmospheric and Planetary Sciences), Daniel S. Kemp (Professor of Chemistry), Kenneth Keniston (Andrew W. Mellon Professor of Human Development, Program in Science, Technology, and Society (STS)), Daniel Kleppner (Lester Wolfe Professor of Physics), Stephen J. Lippard (Professor of Chemistry), Kenneth R. Manning (Thomas Meloy Professor of Rhetoric), Marcia K. McNutt (Professor of Earth, Atmospheric and Planetary Sciences), Richard C. Mulligan (Associate Professor of Biology), Paul Penfield, Jr. (Head, Department of Electrical Engineering and Computer Science), Walter Rosenblith (Institute Professor, Electrical Engineering), Paul R. Schimmel (Professor of Biology), Phillip A. Sharp (Head, Department of Biology), Eugene B. Skolnikoff (Professor of Political Science), Alar Toomre (Professor of Mathematics), Sheila Widnall (Abby Rockefeller Mauze Professor of Aeronautics and Astronautics and Associate Provost), and Gerald N. Wogan (Director, Division of Toxicology). Charles M. Vest (President), Mark S. Wrighton (Provost), John M. Deutch (Karl Taylor Compton Professor of Chemistry), and Joel Moses (Dugald C. Jackson Professor of Computer Science and Engineering and Dean of the School of Engineering) were ex-officio members of the group. David H. Guston (Political Science and STS Program) was the rapporteur.

Judith Stein and Debbie Meinbresse in the office of the STS Program at MIT assisted in managing the workshops, as did the staff of the MIT Faculty Club, where the meetings were held. Wade

Roush served as substitute rapporteur. Judy Spitzer at the STS Program helped transform papers and disks into a manuscript.

The Center for Science and International Affairs at the Kennedy School of Government at Harvard University provided institutional support for David Guston's editorial work, and Teresa Pelton Johnson offered particularly helpful advice.

We are grateful to Louise Goines of the AAAS Press for initial encouragement, and to Larry Cohen of The MIT Press for helping us see the project through.

About the Authors

Daryl E. Chubin has been Division Director for Research, Evaluation, and Dissemination in the Education and Human Resources Directorate of the National Science Foundation since September 1993. Prior to that he was senior associate in the Science, Education, and Transportation Program of the Office of Technology Assessment, U.S. Congress. Dr. Chubin's research has centered on the social and political dimensions of science and technology. He is the author of *Peerless Science: Peer Review and U.S. Science Policy* (SUNY Press, 1990, with E. J. Hackett), and he was project director of the OTA reports *Federally Funded Research: Decisions for a Decade* (1991) and *Educating Scientists and Engineers: Grade School to Grad School* (1988).

David H. Guston is Assistant Professor of Public Policy at the Eagleton Institute of Politics of Rutgers, the State University of New Jersey. His primary research interest is the interaction of democratic institutions and scientific practice, and he has published articles on the congressional oversight of science, both historical and contemporary.

David Hamburg has been President of the Carnegie Corporation of New York since 1983. He is also founder of the Carnegie Commission on Science, Technology and Government. His research has dealt with biological responses and adaptive behavior in stressful circumstances and with aspects of human aggression and conflict resolution. He has also been interested in the conjunction of biomedical and behavioral sciences—first in the context of building an interdisciplinary scientific approach to psychiatric problems, and then in research on the links between behavior and health as a major component in the contemporary burden of illness. He is the author of *Today's Children: Creating a Future for a Generation in Crisis* (Random House, 1992).

Gerald Holton is Mallinckrodt Professor of Physics and Professor of History of Science at Harvard University and Visiting Professor at the Massachusetts Institute of Technology. His books include *Thematic Origins of Scientific Thought: Kepler to Einstein* (Harvard University Press, 1973); *The Scientific Imagination: Case Studies* (Cambridge University Press, 1978); *The Advancement of Science, and Its Burdens: The Jefferson Lecture and Other Essays* (Cambridge University Press, 1986); and *Science and Anti-Science* (Harvard University Press, 1993).

Kenneth Keniston is Andrew W. Mellon Professor of Human Development in the Science, Technology, and Society Program at the Massachusetts Institute of Technology. Before coming to MIT in 1976, Dr. Keniston was the chairperson and director of the Carnegie Council on Children. He is the author of *The Uncommitted* (Harcourt, Brace & World, 1965); *Young Radicals* (Harcourt, Brace & World, 1968); *Youth and Dissent* (Harcourt Brace Jovanovitch, 1971); *Radicals and Militants* (Lexington Books, 1973); and *All Our Children: The American Family Under Pressure* (Harcourt Brace Jovanovich, 1977). His current research focuses on the recruitment, education, and careers of young engineers.

Daniel Kleppner is Lester Wolfe Professor of Physics and Associate Director of the Research Laboratory of Electronics at Massachusetts Institute of Technology. He has served as chairperson of the Division of Atomic, Molecular and Optical Physics of the American Physical Society and is currently the chairperson of the APS Physics Planning Committee. Dr. Kleppner's research interests are in experimental atomic physics, high precision measurements, and quantum optics.

Peter Likins has been President of Lehigh University since 1982. He previously served as Provost of Columbia University and as Dean of the Columbia School of Engineering and Applied Science. Dr. Likins was a member of President Bush's Council of Advisors on Science and Technology, and he is a member of the Business–Higher Education Forum, the executive committee of the Council on Competitiveness, and the Pennsylvania Economic Development Partnership Board.

Dorothy Nelkin is a University Professor at New York University, teaching in the Department of Sociology and the School of Law. Her research focuses on the relationship of science to the public. She has served on the board of directors of the American Association for the Advancement of Science and is currently on the National Advisory Council for Human Genome Research of the National Institutes of Health. Her books include *Science as Intellectual Property* (Macmillan, 1984); *Selling Science: How the Press Covers Science and Technology* (W.H. Freeman, 1987); *Controversy: The Politics of Technical Decisions* (3rd ed., Sage, 1992); and *Dangerous Diagnostics: The Social Power of Biological Information* (University of Chicago Press, 1994, with L. Tancredi).

Phillip A. Sharp was Director of the Center for Cancer Research at the Massachusetts Institute of Technology from 1985 to 1991 and is currently head of MIT's Department of Biology. Dr. Sharp received the 1993 Nobel Prize in Physiology or Medicine for the 1977 discovery of split genes.

Harvey M. Sapolsky is Professor of Public Policy and Organization in the Department of Political Science at the Massachusetts Institute of Technology and Director of both the MIT Defense and Arms Control Studies Program and the MIT Communications Forum. His areas of specialization are science, defense, and health policies. In the science policy field, his most recent work is *Science and the Navy* (Princeton University Press, 1990), a history of the Office of Naval Research. His current research includes studies of the effect that casualties of all types have on U.S. doctrine and actions, the problem of innovation in both civilian and defense policy, and the politics of risk.

Eugene B. Skolnikoff is Professor of Political Science at the Massachusetts Institute of Technology. He served in the Science Advisor's Office in the Eisenhower and Kennedy administrations, and he was a senior consultant to President Carter's Science Advisor. His work in government, research, and teaching has focused especially on the interaction of science and technology with international affairs, covering a wide range of industrial, military, space, economic, environment, and futures issues. His most recent book is *The Elusive Transformation: Science, Technology, and the Evolution of International Politics* (Princeton University Press, 1993).

Albert H. Teich is Director of Science and Policy Programs at the American Association for the Advancement of Science. This directorate is responsible for AAAS programs in science and public policy; science, technology, and society; and science and human rights. Prior to joining the staff at AAAS in 1980, Dr. Teich was an Associate Professor of Public Affairs and Deputy Director of the graduate program in science, technology and public policy at the George Washington University. Among his major publications are *Technology and the Future* (6th ed., St. Martin's Press, 1992), a widely used textbook on technology and society, and *Science and Technology in the U.S.A.*, volume 5 of Longman's *World Guides to Science and Technology* (1986).

Charles M. Vest became the fifteenth President of the Massachusetts Institute of Technology in 1990. Prior to that, he had served as Provost, Vice President for Academic Affairs, and Dean of Engineering at the University of Michigan. Dr. Vest's research interests are in the thermal sciences and the engineering applications of lasers and coherent optics.

Patricia Woolf is currently Lecturer in the Molecular Biology Department at Princeton University. She is also a consultant and author who lectures widely on topics related to scholarly publishing, especially in the sciences. Dr. Woolf is a member of the board of The Scientists' Institute for Public Information and has served on the Panel on Scientific Responsibility and the Conduct of Research of the National Academy of Sciences.

1

Introduction: The Social Contract for Science

David H. Guston and Kenneth Keniston

In the years following World War II, the United States developed a remarkable system for supporting scientific research. This system was founded on a vision of science as an "endless frontier" that could replace the physical frontier of the American West as a driving force for economic growth, rising standards of living, and social change (Bush 1990 [1945]). Scientific discoveries, it was hoped, would not only keep the United States the world's leader in military technology but would also create an endless stream of new commercial products, new medical technologies, and new sources of energy that would eventually benefit all people. The institutions and practices created to support the system were a unique blend of public and private enterprises, eventually including a set of national biomedical laboratories at the National Institutes of Health (NIH), a set of military research and development (R&D) centers such as Los Alamos and Lawrence Livermore National Laboratories, mission agencies with special technological goals such as the National Aeronautics and Space Administration (NASA), and even a National Science Foundation (NSF) to give grants to scientists at public and private research universities.

In many ways, the research universities have been at the intellectual center of this entire enterprise, since it is there that most of the basic science research has been done. At the heart of federal support for universities has been the practice of competitive, peer-reviewed grants. The goal of peer review is simple: Identify the best research as defined by the scientists themselves. And the bargain struck between the federal government and university science—what we call the "social contract for science"—can be summarized in a few words: Government promises to fund the basic science that peer reviewers find most worthy of support, and scientists promise that the research will be performed well and honestly and will provide a steady stream of discoveries that can be translated into new products, medicines, or weapons.

Whether measured in terms of people, products, patents, publications, or prizes, the American system of science has been the most successful in the world. Almost five decades into the social contract for science, however, there are signs that its pattern of partnership and harmony has eroded. It may well be, as the chapters in this volume suggest, that there was never a real "golden age" in the relationship between the federal government and the scientific community. Nevertheless, it is clear that the contract between science and society is undergoing a rethinking such as it has not experienced since its inception.

A Crisis in Science Policy?

In the late 1980s and early 1990s, conflicts between science and government have increased in number and noise. From Washington has come a slew of painful accusations about scientific research in the universities. Congressional committees have investigated cases of alleged scientific fraud and claims that federal funds had been spent by research institutions for liquor, yachts, and even (the supreme irony) lawyers to defend themselves against federal lawsuits. Some members of Congress have argued that the openness of American universities to foreign researchers and students allows our economic competitors to steal scientific and technical secrets whose development has been funded by U.S. taxpayers for the express purpose of competing in the international marketplace.

Attempts have been made to portray scientists (or at least the institutions in which they work) as generally greedy and selfish in their unending quest for new funds, as witnessed by their unwillingness to set priorities and their constant complaining when requests for funding are denied. University scientists are also attacked for supposedly neglecting teaching and research in order to enrich themselves through consulting relationships and spinoff corporations. Throughout all of these accusations runs the implication that academic scientists have become arrogant and self-indulgent, rejecting legitimate oversight of the use of public money, claiming "entitlement" to ever-escalating funding, and unwilling to share responsibility for dealing with the growing deficits, trade imbalances, and other economic ills of their country.

The complaints voiced by the scientific community about government are scarcely less vehement. Congress and the executive branch stand accused of intruding into the conduct of science itself, attempting to "micromanage" scientific investigations, confusing honest mistakes with fraud, and subjecting distinguished scientists to humiliating and often ignorant cross-examination.

Far from being overindulged, many scientists claim, they are underfunded. A smaller proportion of research proposals are now being approved than in the past; outstanding researchers must waste time applying for multiple grants because so many requests to federal funding agencies are refused. Far from supporting luxuries and frills, federal grants for the indirect costs of scientific research do not even provide adequate compensation for the basic costs of running a research institution. Moreover, federal funding of research is often so delayed, or so laced with constraints, that responsible financial planning has become increasingly difficult for the research universities.

Other scientists believe that the growing congressional practice of "earmarking" R&D funds (specifying the precise institution or region to which funding should be given) is undermining the entire system of merit-based, peer-reviewed support that has made American science the envy of the world. And not least of all, academic administrators complain that onerous reporting requirements imposed upon applicants by the federal government require vast, expensive, and unproductive administrative staffs to assure that every requirement, no matter how trivial or unreasonable, is fully complied with.

The current situation in science and technology policy thus shows some signs of a conflict in which each side publicly attacks the motives of the other and expresses fear for the continuity of its own values. Indicative of this atmosphere of apparent crisis is the sheer volume of printed analyses, reports, recommendations, and suggestions (a sampling would include OTA 1991; Carnegie Commission 1992; GUIRR 1992; NSB 1992; and U.S. Congress 1992).[1]

Despite all these analyses, the underlying causes of the current conflict between government and science have not been evident, nor have definitive assessments of its significance been made. Is an increasingly selfish and arrogant scientific community to blame? Or have politicians bent on personal aggrandizement torn down science in order to build themselves up? Are the recent controversies simply passing waves on the always tumultuous sea of public policy, or are they reflections of new political trends, new scientific directions, and newly emerging structures? Has the maw of the federal budget deficit been devouring science and technology funding, or is science peculiar in its need for more funds and ever-larger projects?

In short, is there a "crisis" in science and technology policy in the United States?

Most recent public discussion has focused on the more spectacular controversies of the past few years. In this book, we aim instead at a middle level of analysis, moving away from specific cases to address more general questions about the nature of the current relationship between politics and science. Our goal is to clarify both the constant elements and the new variables in that relationship.

Science practice and science policy are, of course, part of a larger social and political context, and controversies within science policy are inevitably linked to broader national trends and controversies. In the end, we do not believe that the current controversies between science and politics are indicative of a new or terminal "crisis" in science policy, at least not if "crisis" implies a discontinuous transition from something familiar into something unrecognizable. In this introduction, we argue instead that, given the inevitable stresses on political institutions and on the research enterprise, what keeps the turmoil from becoming a true crisis is the continuing contract between science and society, however fragile it may be.

The Social Contract between Government and Science

A Useful Metaphor

The idea of a social contract is commonly invoked to describe the relations between the communities of science and government. Sometimes termed a "tacit contract" or a "social contract for science," this metaphor is rooted in the actual contracts that establish the relationships between the federal government and scientists (see, e.g., Price 1954) as well as in the metaphorical contracts that bind and unite professional communities like that of scientists. In the language of science policy, the "social contract for science" refers above all to the constitution of the post–World War II research system on the blueprint outlined by two reports: Vannevar Bush's *Science: The Endless Frontier* (1990 [1945]) and John R. Steelman's *Science and Public Policy* (1947).

The metaphor of the contract is useful for several reasons. A contract implies two distinct parties, each with different interests, who come together to reach a formal agreement on some common goal. Implicit, too, is the notion that a contract is negotiated, arrived at through a series of exchanges in which each party tries to secure the most advantageous terms. A contract, moreover, suggests the possibility of conflict—or at least disparity of interests. For example, we do not usually make contracts with ourselves or with our immediate family; when we do, as with prenuptial agreements, they acknowledge the possibility of potential future conflicts. Finally, contracts can be renegotiated if conditions change for either party.

In contemporary usage, the contract metaphor also suggests the privileged treatment of the science community by government. Representative George Brown (D-CA) writes:

> Science and the technology that it spawns are viewed as a cornerstone of our past, the strength of our present, and the hope of our future. An unofficial contract between the scientific community and society has arisen from these beliefs. This contract confers special privileges and freedoms on scientists, in the expectation that they will deliver great benefits to society as a whole. (1992: 781)

But as Brown makes clear, this contract may need replacing or renewing: "The scientific community must seek to establish a new contract with policy makers, based not on demands for autonomy and ever increasing budgets, but on the implementation of an explicit research agenda rooted in [social] goals" (1992: 780).

Science as a Public Good

What are the basic elements of the contract between science and government? First, it involves the provision of a public good, namely useful research, by the scientific community. The public good argument—that the private sector will underinvest in scientific research because it is difficult for private patrons to appropriate the return on their investment—has been invoked in the science policy debate in the United States since the founding of the Republic. The protection of patents and copyrights in the U.S. Constitution (Article I, section 8) is an attempt to encourage invention by offering protective monopolies to creators and inventors. By the mid- to late nineteenth century, the relative lack of private patronage of science was part of the rationale for an expanding federal role in funding research in the federal bureaus and in the nascent universities (Bruce 1987; Turner 1990). Although corporations and philanthropies began to invest larger amounts in scientific research by the turn of the century and particularly after World War I, much of this funding dried up during the Great Depression, re-emphasizing the unreliability of private support (Weiner 1970; Kevles 1978, 1992).

Throughout the post–World War II era, advocates of federal support for basic scientific research have underscored the federal obligation to provide this public good. Economists distinguish between public "consumptive" goods and public "productive" goods (Rottenberg 1968). As a consumptive good, science is an investment for culture akin to other public investments in parks, museums, and the arts. But in part because of the ideology of limited government and the greater expense of science relative to the arts, the continued expansion of government support has been justified above all with claims that science is uniquely useful as a productive good (Lederman 1991; Mansfield 1991; CBO 1993).

The logic of the public good aspect of the social contract for science extends to technology as well. For example, arguments attempting to justify a federal role in the creation and dissemination of new technologies use the public good argument to expand it but also to constrain it: Some investments in technology are difficult for a single firm to appropriate, so there is a role for the government; but the governmental role should be limited to technologies that cannot be appropriated by the private sector (Alic et al. 1992; Bromley 1993).

In the years of the Cold War, the role of science in strengthening the defenses of the United States and the "free world" was especially important in generating public support for the scientific community. Perhaps the single most important agency supporting basic research in the universities after World War II was the Office of Naval Research (ONR) (see Sapolsky 1990). The level of funding of the NSF did not become significant until after the Soviets launched their Sputnik satellite in 1957. The impetus given by Sputnik to NSF funding demonstrated the extent to which the hope for military benefits lay behind government support of basic research even in areas not immediately related to military applications.[2]

Just as the development of radar, the proximity fuse, and the atomic bomb during World War II revolutionized warfare, so too the wartime development of blood plasma and antibiotics revolutionized the treatment of injuries and infectious diseases, giving credibility to the argument that further research would result in cures of even the more intractable conditions such as cancer, stroke, heart disease, and in recent years AIDS. NIH was transformed from a sleepy laboratory into "a brilliant jewel in the crown" of the federal research establishment (Harden 1986: 179), providing some $5.7 billion for basic biomedical research in 1993 alone. This sum amounts to more than 42 percent of the entire federal portfolio of basic research and exceeds the combined basic research budgets of NSF, NASA, and the Department of Energy (DOE) (AAAS 1994).

Public investment in biomedical research illustrates another characteristic of the "investment" rationale for the support of science: It involves one generation investing for the benefit of the next. For example, it is assumed that funds invested today in fundamental problems of molecular biology will aid future generations, even if specific discoveries and cures arrive too late to benefit those whose tax dollars funded the research. In a similar way, of course, future generations must bear the burden of unwise scientific research or technologies that have harmful side effects. Especially in an era of budgetary strains and widespread concern for the long-term consequences of technological change, defining and achieving generational equity in science funding and technology development is a vital question.

Accountability and Autonomy

A second major characteristic of the social contract for science is the special mechanisms by which it balances responsibilities between government and science. This balance takes into consideration both the values of accountability associated with representative government and those of autonomy associated with an independent professional community. Not only does the government "invest" in a public good (as it does to provide for the common defense), but it delegates to other institutions the actual conduct of the research.[3] It is thus the scientific community, as established in universities and other research institutions, that has responsibility for "producing" research, discoveries, and new technologies. But the government's delegation of research is neither a gift without strings nor a procurement for specific products. It is instead a set of specific grants and contracts, each with its own terms and conditions.[4]

The federal funding agencies generally make use of some form of peer review, in which scientists rate proposals for specific projects on a competitive basis. Peer review varies by agency and program as to whether the scientists who do the reviewing are actually federal employees of the agency or whether they are extramural scientists engaged for the sole purpose of review. Even in the paradigmatic cases of NSF and NIH peer review, the conclusions of the review committees are recommendations only, although in practice they are rarely overruled by administrators.[5] This contrasts with governments, such as France, that make block grants to research institutions, which in turn use the funds for whatever purposes they deem important.

The invention of the peer review system during and after World War II instituted a new type of federalism that attempted to perpetuate two apparently incompatible goals: the government would sponsor research and to a large degree designate the broad areas to which research support would go; but the scientific community and its institutions would remain independent and responsible for their own affairs (Price 1954). The social contract for science has sought to further both of these goals.

From the government's point of view, it has proven exceedingly difficult to know whether or not scientists are keeping their side of the bargain. Philosopher of science Stephen Turner (1990: 189) notes that "the general 'contract' between the scientific community and its primary patron, the state, and the individual relations of patronage embodied, for example, in particular grants or particular bureaucrats' personal relations to scientists, are both uncheckable." By "uncheckable" Turner means that the terms of the contract make it impossible to verify the integrity and productivity of research in a manner agreeable to both parties.

Conversely, it is difficult for the scientific community to guarantee that government support will be handed out on a "fair" basis, that is, on the basis of merit alone. As noted above, the organized scientific community publicly deplores any move away from peer review and toward earmarking, even though individual scientists or research institutions are often behind the contact with politicians that leads to earmarking.

Having realized that the social contract for science is "uncheckable" and worried that the scientific community is no longer reliable, politicians have begun to insist on measuring the integrity and productivity of science in a more direct way and on increasing political participation in promoting these goals (Guston 1993c). But as the fragile contract becomes more explicit, it becomes harder to maintain. Once the measuring of scientific integrity and productivity begins, every case of scientific misconduct and every patent taken out on a product of sponsored research becomes a potential cause for renegotiation by congressional overseers.

The Social Contract for Scientists

The metaphor of a social contract is also useful in describing relations within the scientific community. Sociologists and historians have long described something like a social contract under which scientists agree to obey implicit rules in the production of knowledge, rules such as accurately and truthfully reporting observations or acknowledging indebtedness to the ideas of others (Polanyi 1962, 1964 [1946]; Zuckerman 1977, 1984). As scientist and philosopher Michael Polanyi puts it, the social contract among scientists constitutes a "Republic of Science," which "realizes the ideal of Rousseau, of a community in which each is equal partner in a General Will" (Polanyi 1964 [1946]: 16). As citizens of this ideal republic, scientists submit to the general will as represented by scientific opinion. "This absolute submission leaves each free but each is also obligated and devoted to the ideals of scientific work" (Polanyi 1964 [1946]: 64).

As sociologist of science Harriet Zuckerman (1977: 113) explains, "an implicit social contract [among] scientists" is a specific example of other "tacit social contract[s] among professionals" that justify the norms of professional behavior. This idea of the social contract for scientists is at the heart of scientific self-regulation, in which the republic of science is thought to permit some scientists to make errors (or even to commit fraud) because other scientists are "obligated and devoted" to checking data, correcting mistakes, and replicating and extending experiments (see, e.g., Taubes 1993).

Critics of contemporary science sometimes doubt whether the vague notion of an implicit social contract is enough to guarantee the professional integrity of science.[6] Some argue that more formal mechanisms of oversight, especially of the integrity and accuracy of scientific results, are needed.[7] But the idea of a tacit agreement among scientists is often voiced within the scientific community and is largely taken for granted in the socialization of researchers. Here, for example, is a statement from a style manual for publishing in biomedical research:

> Scientists build their concepts and theories with individual bricks of scientifically ascertained facts, found by themselves and their predeces-

sors. Scientists can proceed with confidence only if they can assume the previously reported facts on which their work is based are indeed correct. Thus all scientists have an unwritten contract with their contemporaries and those whose work will follow to provide observations honestly obtained, recorded, and published. (CBE 1983: 1)[8]

In this sense, there is not only an "unwritten contract" among scientists, but a written one as well. The research paper that bears the names of each of those responsible for the observations and theories detailed in it is a public witness to the understanding described above. Placing one's name on a scientific paper without ascertaining the integrity of its contents is like signing a contract without understanding the duties and obligations enumerated within it. In many cases, the expectation is made explicit: for example, *The New England Journal of Medicine* requires each of the (frequently many) authors of papers published in its pages to attest separately to the accuracy of all of the findings reported in the paper.

Consensus and Change

The idea of a social contract for science that emerged and flourished in the United States in the years following World War II has yet another use. It expresses an original consensus against which change can be measured and evaluated. If we are correct in identifying a set of shared assumptions and agreements on which the postwar relationship between the political and the scientific communities was premised, we can evaluate the extent to which those premises have changed over time, have been contested, or have been proven false.

Harvey Brooks, one of the original elaborators of the social contract metaphor, has argued that:

Many times in the last 45 years commentators have predicted the imminent dissolution of the "social contract" between the scientific community and the polity that was so cogently formulated in the Bush report. The year 1971 was a particularly acute time of "doom and gloom" in this respect, and yet the relationship has not deteriorated to the extent anticipated. Today, many of the same negative signals that existed in 1971 are again evident. Will science recover to experience a new era of prosperity as it did beginning in the late 70s, or has the day of reckoning that so many predicted finally arrived? Only time will tell. (Brooks 1990: 33)

More recently, however, Brooks (1993) has acknowledged changes in the social contract for science.

A few commentators are dubious about the idea that there was a clear contract in the postwar era. For example, political scientist Richard Barke (1990: 1) argues that "allegations of reneging on the `compact' between science, technology, and government are often based on rather selective interpretation of Vannevar Bush's message . . . ; yet some fundamental changes do appear to be emerging in the relationship between the scientific enterprise and its public patron." But most others, especially those who speak for the scientific community, accept the notion of a contract without reserve and defend the community's faithfulness to it. For example, Frank Press, the former president of the National Academy of Sciences, asserts that "Science has been faithful to that compact—that the American people for their support of science could in time expect a better life and a stronger nation. And we continue to honor that compact" (1988: 2).

Our own reading suggests the usefulness and validity of the idea of a postwar social contract between the federal government and the scientific community that led to a uniquely American system of basic scientific research. The question the authors of this volume raise is whether this contract—universally perceived as more fragile today than in 1950—can long endure.

The Golden Age

There is probably a universal human tendency to simplify and idealize the past. In the case of science policy, some scientists adhere to a myth of the "golden age" in which money flowed freely—as needed—from Washington to research institutions, in which scientists themselves (through peer-review panels established within the federal government) decided on the most meritorious projects to be funded, and in which the scientific community could count on steady increments in federal support every year. During these years, furthermore, it is said that government did not "interfere" with scientists, who guided their own affairs according to their own lights. Basic research was guided solely by the interests of scientist and the intellectual dynamics of research fields; crass pressures for immediate usefulness or applicability were almost entirely absent. In a word, the federal government provided the money; science provided the discoveries and kept its own house in order.

Since current tensions between government and science are often compared (explicitly or implicitly) with this picture of the past, it is important to ask how accurate it is. Was the relation between government and science ever as harmonious as this? Was the scientific community ever completely free from government interference? Was the pressure toward commercial, industrial, and other forms of utility ever entirely absent? Was the scientific community in fact considered accountable only to itself, and not to the public agencies that were funding it?

In this regard, the definitive history of science policy in the United States remains to be written.[9] Here we can only touch on issues and episodes that, in our view, muddle any rosy picture of a golden age.

By the early 1950s, the basic institutions of federal science—ONR, NIH, NSF and the Atomic Energy Commission (AEC)—were well-established and already held in high regard, but the relationship between government and science was not untroubled. Even as the funding agencies were establishing their programs and peer-review systems, questions about civilian control of atomic research and about the loyalty of scientists were challenging the new relationship. For example, the National Science Foundation Act that President Truman signed into law in 1950 specified that grant applicants had to sign an affidavit of loyalty. Both the legislative and executive branches intervened to declare certain scientists disloyal and therefore ineligible for research grants and positions of public responsibility in areas even remotely related to national security. The Department of Health, Education, and Welfare, the parent department of NIH, suspended funding for a number of scientists for "engaging in subversive activities" or for being subject to "serious questions of . . . loyalty to the United States" (Edsall 1988 [1955]: 24). Although some members of the scientific community and some of its more outspoken organizations expressed strong opposition, the community at large did not rise to the defense of those attacked (Wang 1992).

These affidavits, security hearings, and loyalty checks demonstrated that politicians were quite capable, if they desired, of overseeing scientists in a minute way, despite the more general claim that scientists should run their own affairs. Loyalty issues further illustrated the fragility of the grant and contract system that was intended to preserve the independence of research institutions: Because they are the legal recipients of federal funding, universities became the (sometimes reluctant) agents of federal policy as they were asked to enforce the loyalty requirements attached to grants and contracts.

The government's insistence on financial accountability also has a long history. For example, in the early 1960s, Representative Lawrence Fountain (D-NC) inquired into the fiscal integrity of grant and contract programs at the burgeoning NIH (Strickland 1972) and eventually managed to force NIH to tighten some of its accounting procedures. Similarly, early negotiations between funding agencies and research universities about indirect costs, while less acrimonious and less public than current debates, established this issue as one of constant renegotiation and bargaining between the universities and the government from the 1950s to the present. It was never the case that the federal government simply handed over the money and asked no questions.

Furthermore, even in the 1960s and 1970s, questions of the social productivity and relevance of science increasingly attracted congressional attention. During this era, Congress instituted the Research Applied to National Needs (RANN) Program at NSF, the War on Cancer at NIH, and the Mansfield Amendment for defense research.[10] In each case, programmatic directions for research originated not from scientists but from Congress, which had in mind social goals not necessarily aligned with those of the scientists. These new programs resulted in changes in program administration and direction that increased at least the appearance of the accountability of scientists. RANN and the War on Cancer also made accountability for productivity a positive-sum game—a game in which everyone won. Scientists accepted more programmatic direction in exchange for more money, while Congress answered potential critics of research funding by insisting on pertinent, relevant, and useful projects. This same tradeoff has continued in such contemporary projects as women's health programs, AIDS research, and global change research.

The experience with federally funded research designed to produce specific results suggests a rule applicable to the present. As the level of domestic spending stabilized and in some cases contracted through the 1980s, members of Congress began to oversee programs with far more exacting scrutiny than before. Even though scientific budgets were largely immune to the contraction suffered by domestic programs in the 1980s—suggesting that science remained in some sense privileged—congressional scrutiny of science increased as well, as if to determine whether that immunity was warranted. It is likely that as budgets become tighter still, the scrutiny will become even more exacting. Given the stringent budgetary realities of the 1990s, politicians who want to support spending for science can help sustain it by articulating—and encouraging scientists to articulate—clear programmatic directions.

Perhaps the most consistent source of tension in the government-science contract has been the funding of defense-related research. Much of the impetus for continuing generous federal funding of science after World War II grew out of the military successes of research-based inventions. The Cold War made it imperative in the eyes of Washington not only to continue research and development of weapons systems but also to encourage basic research in areas that might ultimately prove militarily useful. To Pentagon planners who anticipated that future wars would be won or lost by technological superiority, funding scientists in universities was seen as a relatively inexpensive way to maintain the technological and intellectual bases of military readiness (Mukerji 1989). Defense-related research institutions were created to conduct classified research under the aegis of major research universities: for example, MIT's Instrumentation Laboratory, which developed missile guidance systems; the Lawrence Livermore and Los Alamos National Laboratories, administered by the University of California, which developed nuclear weapons; and the Jet Propulsion

Laboratory at the California Institute of Technology. These laboratories also provided indirect support for many scientific projects at the universities and training and employment for many graduate students and faculty.

In the eyes of many research scientists and others, however, the secrecy necessary in military research violated a cardinal principle of science, namely, the open publication of results (Foerstel 1993). Moreover, many researchers, students, and administrators objected to the presence on campus, or under the auspices of the university, of research projects whose purpose was to create instruments of death and mass destruction. Especially after concern about the war in Vietnam escalated in the late 1960s, pressures to dissociate universities from all classified (secret) research led to conflicts both within the universities and between universities and the federal government (see, e.g., Allen 1970; Nelkin 1972). The more recent rejection by many scientists of funding associated with the Strategic Defense Initiative (SDI, or "Star Wars") recalled earlier debates that had raged around other weapon and antiweapon systems (see Primack and von Hippel 1974).

But even though the tensions between the scientific community and the military have been many, the existence of important military needs has been of unquestionable benefit to the scientific community. In all but a handful of years since 1945, the military has provided more than 50 percent of federal R&D expenditures. Equally important for basic science, the general enthusiasm for science that emerged from World War II supported not only specific military projects but also more diffuse research financing in physics, nuclear engineering, oceanography, and, later, mathematics, electrical engineering, computer science, and artificial intelligence.

The hope—and to a degree the reality—was that support of basic science by the federal government could be the wellspring of a scientific and technological community capable of creating a decisive technological edge over the Soviets. In fact, some argue that the scientific and technological backwardness of the Soviet Union vis-à-vis the United States was a major reason for the internal reforms instituted by Mikhail Gorbachev, and thus that American science ultimately contributed to the collapse of the Soviet Union and the end of the Cold War.

In short, most if not all of the current conflicts between government and science have roots in the beginnings of their relationship in the immediate postwar period. Congressional inquiries, the insistence on relevance, the use of grants and contracts as instruments of policy—all have a long history. So too, as discussed below, do budgetary difficulties. Because of these historical realities, much contemporary rhetoric about a crisis in science policy based on new and unprecedented political pressures is misplaced.

Changed Government, Changed Science

The golden age was never quite so golden as it seems to some nostalgic scientists. But it is also true that much has changed over the decades since the social contract for science was established, and that these changes are largely irreversible. Indeed, perhaps the simplest explanation for the current heightened tensions between government and science is that the old contract was made between a kind of government that no longer exists and a kind of scientific community that has long since disappeared.

In the postwar years, both the executive and the legislative branches have changed in ways relevant to the governance of science. At the executive level, the postwar years witnessed the rise of an "imperial presidency," which extended presidential prerogatives far beyond their prewar limits, and a "management presidency," based in the Office of Management and Budget (OMB), which tried to coordinate the expanding bureaucracy. The White House added analytical capabilities with the Special Assistant to the President for Science and Technology and the President's Science Advisory Committee (PSAC) in 1957; the Office of Science and Technology (OST), which in 1976 became the Office of Science and Technology Policy (OSTP); and the Federal Coordinating Council on Science, Engineering, and Technology (FCCSET), which in 1993 was re-created as the National Science and Technology Council (NSTC).[11] Scientific advisory committees also proliferated in other departments and agencies of the executive branch.[12]

On the congressional side, the power of committee chairmen declined through the postwar years and has been replaced by a more decentralized organization, characterized by greater participation from subcommittees as well as action outside of committees.

There has also been a general resurgence of congressional oversight directed at maintaining accountability over the burgeoning programs and agencies of the executive branch. Congressional staffs have increased in number and professional competence, and Congress has augmented its analytical capabilities through the creation of the Office of Technology Assessment (OTA) and the Congressional Budget Office (CBO), the expansion of the Congressional Research Service (CRS), and the increasing control over the General Accounting Office (GAO). Congress has also created Offices of Inspectors General in all the major departments and agencies to monitor the execution and implementation of policy.[13]

No one would claim that these larger changes were made primarily because of their impact on the system of science and technology. Even the creation of OTA was mostly a result of institutional conflict between the President and Congress, rather than the result of a plan (although a plan existed) to improve science and technology policy by forecasting its likely social impacts (Bimber 1992). In very general terms, however, these changes have tended to give both branches increased competence and motivation to oversee and evaluate the scientific community.

In the context of these larger institutional changes, the pattern of federal funding for R&D is also much more complicated than the fond recollections of a golden age gone away. Devotees of the golden age selectively recall the mid-1960s, when federal R&D spending reached an all-time high. In terms of percentage of the gross national product (GNP), 1964 was a maximum, with the federal government expending almost 2 percent of the GNP for R&D. In terms of federal R&D as a share of federal spending, the peak came in 1965; in terms of real spending, the peak was 1966 and 1967 (see table 1.1).

It must be noted, however, that this mid-1960s peak was inordinately inflated by the Apollo program and by the heat of global competition with the Soviets: Over the decade from 1963 to 1972, national defense R&D accounted for almost 54 percent of federal expenditures; space accounted for another 27 percent (NSB 1973: tables 16 and 17). Today, with the dissolution of the Soviet threat and the absence of a coherent and exciting space program, there are no comparable goals for basic scientific research that have a

Table 1.1 Federal R&D expenditures, variously expressed

	% of GNP	% of federal outlays	constant 1958 dollars (billions)
1964	**1.985**	12.3	11.5
1965	1.903	**12.4**	11.8
1966	1.866	11.8	**12.3**
1967	1.816	10.7	**12.3**

Peak years are in boldface. Source: NSB (1973).

strong national consensus. Although the Reagan defense build-up raised average defense R&D spending over the decade from 1983 to 1992 to about 56 percent of total federal R&D (NSB 1991), the defense share in years since 1992 has fallen to less than 54 percent (AAAS 1994). Furthermore, space-related R&D accounted for only 7 percent of total federal R&D during the 1980s (NSB 1991). Only in the general area of health does a ready consensus exist for massive public spending on research, and even there it is often riven by competition among research priorities, for example AIDS vs. mental health vs. women's health research.

Although some decrease in federal R&D spending is indisputable, it is not clear just how far from the 1960s peaks the level of spending has fallen. Comparing expenditures from one year to the next in a meaningful way requires taking inflation into account, but there is no standard deflator to apply to R&D spending: NSF uses one deflator and OMB uses another. Using the NSF deflator, spending in 1990 in constant dollars was about 30 percent *higher* than the 1966 peak; using the OMB deflator, spending in 1990 remained below that peak (OTA 1991: box 2-C).

Another way to look at R&D spending is in the context of the rest of the federal budget. Over the past decade the share of R&D in the domestic discretionary budget has risen while almost all other domestic discretionary items have fallen. That is, through the 1980s, R&D consumed an increasingly larger share of the increasingly smaller pie of nondefense, nonentitlement spending. For this reason, calls for greatly increased science budgets, such as that made by Leon Lederman (1991) in *Science: The End of the Frontier?*

were probably ill-starred from the beginning. Both the Bush and Clinton administrations have been committed in principle to large increases in R&D spending, but in fact there have only been a few significant increases at NSF and marginal increases at NIH. The sufferings of scientists may be real, but in the words of one of the strongest congressional patrons of science, they are not unique (Brown 1991).

If government has changed in the past five decades, so has science. The scientific enterprise has grown vastly in workforce, complexity, size of projects, and costliness. The growth in the scientific workforce alone is staggering. In 1965 the number of scientists and engineers in the United States was 495,000; by 1988 it had risen to about 950,000.[14] The proportion of the workforce of scientists and engineers engaged in R&D, after shrinking from its previous high of 67.9 per 10,000 in 1968, has increased steadily since 1977, and in 1987 reached 75.9 per 10,000, a 12 percent increase (NSB 1991: 300–301). As historian and physicist Derek de Solla Price (1986) argued, the exponential growth that characterized science through the modern era could not be sustained forever. To take an extreme example, if each of 10,000 research scientists trained only 10 research scientists in his or her lifetime, within five generations there would be one billion research scientists!

Nevertheless, one goal of the federal funding of research has been to increase the number of Ph.D.s so as to provide the nation with a highly trained scientific workforce. However commendable this goal, it has a bizarre consequence: The more successful the program is in creating Ph.D.s, the greater will be the future demand for research financing. It is rather as if a welfare program created a half-dozen new welfare recipients for every one who was taken off federal assistance.[15] This large increase in the number of scientists (and the commensurate increase in the number of research laboratories and universities engaged in basic research) means that, despite real increases in R&D funding, a smaller percentage of grant applications can be funded each year. Principal investigators must therefore, as they frequently complain, file multiple applications to attempt to win adequate support. The scientific community feels that there is less and less money, and per capita the scarcity is real. Surely some truly meritorious research projects today go unfunded; but even if all were funded today, they could not all be funded tomorrow.

Science has also changed in the increasing size, complexity, and expense of its projects. The Manhattan Project and other wartime science inaugurated a trend toward what has been called "megascience." Scientific research in general has gotten bigger and costlier, involving more people on each project and requiring ever more expensive equipment. Science today is a complex aggregate of complex goods such as new technologies and education; as a result, the "research price index" goes up much faster than inflation.

One result of this increase in size and complexity is that more money supports fewer projects. From the point of view of the granting agencies, the government is supporting science more generously then ever before; from the point of view of the individual scientist or project director, grants are increasingly hard to obtain. Many of the contemporary tensions between government and science spring from this difference in perspective. Government officials reasonably and accurately point to their increasing largesse; scientists reasonably and accurately point to the increasing difficulty in obtaining support.

Because even the routine science done in traditional research laboratories has increased in size and complexity, the truly gigantic "megaprojects" may actually be shrinking in attractiveness and plausibility. The space station Freedom, saved from complete budgetary destruction by virtue of its geographically well-distributed construction and R&D contracts,[16] is increasingly reduced in size and viewed as a joint technology development project with the Russians. The Human Genome Project, if it ever was megascience, has been dismembered and redistributed to individual investigators, much as the War on Cancer had been in its turn. In late 1993, Congress terminated the Texas-based Superconducting Supercollider (SSC) project, partly because its practical utility to the nation was in question, but also because the benefits of its construction and its R&D contracts were not as broadly distributed as those of the space station and because the influence of Texans had diminished in both the White House and on Capitol Hill.[17]

But of all the changes since the postwar negotiation of the social contract for science, the end of the Cold War may prove the most consequential. Since 1945 the promise of military applications and the specter of Soviet competition has driven federal R&D expenditures in both military and civilian agencies. More generally, the instrumentality of R&D to the conduct of the Cold War—both in material terms of accurate missiles and in symbolic terms of conquering the new frontier of outer space—meant that governments (in the former Soviet Union as well as in the United States) looked at science as a whole in a favorable light. But without an implacable communist menace, the instrumental argument for science and technology has today lost much of its force.

Of course, the end of the Cold War does not mean the end of war. Especially in the wake of what was perceived to be a technological triumph in the Gulf War, the Pentagon has been quick to underline the need for continuing military R&D to maintain American technological superiority. But the downsizing of the military, the loss of mission of the major weapons laboratories, and the dismantling of SDI are all producing a major reorientation.

The result, especially for the physical sciences, is that a new rationale for public support of science is needed. Previously, one central "public good" upon which almost everyone agreed was countering the Soviet threat. Today, other public goods are alleged—or more precisely revived with new vehemence from the past. For Vannevar Bush, John Steelman, and others who established modern U.S. science policy in the 1940s, the military rationale was only one among many; the rationale of human betterment through increased employment, a rising standard of living, and better health was equally significant. The health claim never lost its persuasiveness, but the rationales of employment and living standards are now being revived and redefined in contemporary terms.

That redefinition mostly takes the form of the claim that science-based innovation is the essential ingredient of the elixir that will revive the nation's economy and restore it to an internationally competitive position. According to this argument, science-based innovations, which have already produced entire new industries such as biotechnology, are constantly needed to maintain the a technological advantage in the international competition for markets and high-wage jobs. In its simplest form, the argument states that there is a direct link between the advances of science in a nation, national success in the international marketplace, and the standard of living in that nation.

In this simple formula, the argument is open to obvious criticisms. The United States is unquestionably the world's leading scientific power, but it lags by international standards in health, has faltered in productivity gains, and is being overtaken by other nations in standard of living. More sophisticated versions of the argument therefore tend to prevail in policy circles: Good science is a necessary but not sufficient condition for productivity (see, e.g., Dertouzos et al. 1989; Branscomb 1993). Moreover, a primary point of this more subtle formulation maintains that the organization of research as established in the postwar years, regardless of how successful it has been to date, is inappropriate for the new environment because it was geared toward a different set of political, technological, and economic challenges.

Meanwhile, public support for science has waned. The almost unqualified public enthusiasm for science and technology that characterized the immediate postwar period has given way to a far more nuanced view. Especially since the 1960s, science and technology have become the objects of extensive public scrutiny and pointed cultural criticism, a criticism by no means limited to cultural dissidents. It was, of course, President Dwight D. Eisenhower who heralded the halcyon days of science advising by appointing James R. Killian, Jr., as the first Special Assistant to the President for Science and Technology and by creating the PSAC.[18] But it was the same Eisenhower who warned in his farewell address that "public policy could itself become the captive of a scientific-technological elite" (1961: 1039).

Eisenhower's misgivings about a "scientific-technological elite" found expression through a number of voices in the 1960s, among them social critics like Theodore Roszak, environmental activists like Rachel Carson, and antimilitary activists such as the March 4th Movement at MIT. In its most limited form, this criticism of science focused on its involvement with the military. But deeper worries about the impacts and governance of science have also spread through both lay and scientific publics. Indeed, scientists themselves have sometimes accused their colleagues of serving vested interests and have called for more "science for the people" (Primack and von Hippel 1974).

Another threat to the close identification of science with the public interest has been the spectacular failure of highly visible technological systems. Dramatic technological failures such as Three Mile Island, Chernobyl, and Challenger may indeed be inevitable in the normal course of events (Perrow 1984). But beyond the immediate losses they cause, disasters also serve as reminders of the potentially harmful impacts of technological change in general (Roush 1993). They suggest that promises about the inevitable utility of scientific research may have been false, and they raise questions about the truthfulness of the scientific and technological elites. Although the public remains hopeful about the possibilities of scientific and technological progress, it is increasingly aware that even the brightest new technology carries possibly harmful consequences. Political scientist Langdon Winner (1986: 6) has written that when we create technological innovations without fully understanding their significance or consequences, "we repeatedly enter into a series of social contracts, the terms of which are revealed only after the signing." Technological disasters remind us how fragile these contracts are.[19]

Even this brief and admittedly superficial analysis of changes in the past five decades suggests that the so-called crisis in government-science relations was inevitable and even necessary. Government has increased in size, complexity, and its capacity both to support and to oversee science. Science, too, has grown from a small business to a multibillion dollar enterprise of extraordinary complexity that has links to every aspect of American society. The original "social contract" was written in simpler days. It is fragile today because the two parties that agreed to it have grown enormously and changed in nature. The tensions and conflicts that became more visible in the 1980s and early 1990s are in good part consequences of these changes.

The Tensions between Science and Democracy[20]

Imagine a select group of members of Congress asking for a special report from the National Academy of Sciences, visiting research centers, and taking testimony from top scientists on: priorities in science funding, the role of different sectors and institutions in the scientific enterprise, the tension between centralization and pluralism in research, the relative merits of large-scale versus small-scale projects, and cost accounting and the fiscal accountability of researchers. Is this Representative George Brown's recent Task Force on the Health of Research? The Science Policy Task Force of the 99th Congress under Representative Don Fuqua? The Fountain Committee, the Elliot Committee, or the Daddario Subcommittee of the 1960s? The Joint Committee on Atomic Energy in the 1950s? The Kilgore Committee at the end of the war?[21]

No, it is the Allison Commission of the 1880s, a select congressional committee that examined all these questions and more with regard to the federal scientific establishment.[22] Like a dysfunctional family, the science policy community in the United States seems to confront the same problems over and over again, never resolving them, no matter how often they arise. Why is it that the same problems arise and that no institutional arrangements seem capable of eliminating the tensions between government and science?

One could look for an answer in the complaints that these communities make about each other, but these often have a partisan tone: lack of knowledge of, appreciation for, and sensitivity toward the necessities of the scientific enterprise are attributed to politicians; and arrogance, elitism, and political naiveté are attributed to scientists. But the similarity of the clash a century ago between Representative Hilary Herbert (D-AL) of the Allison Commission and John Wesley Powell of the U.S. Geological Survey to the recent confrontation between Representative John Dingell (D-MI) and president Donald Kennedy of Stanford University exists not because politicians are ignorant or scientists arrogant. Rather, the basic reason for the recurring tensions between the political and scientific communities lies in fundamental differences between the principles of organization of a democratic polity and those of the scientific community.

There are three principal categories of tensions between democratic politics and scientific practice. Each has direct policy implications. We might call these the populist tension, the plutocratic tension, and the exclusionary tension (Guston 1993b).

The populist tension reflects the fact that popular tastes and preferences are different from, and sometimes antagonistic toward, those of the scientific community. Such antagonism might result, for example, in popular pressure for a more equitable geographic distribution of research funds, for more applied research, for a particular research focus, or for a greater emphasis on teaching or patenting than on research. The populist tension exists because democratic politics expresses, and democratic institutions are controlled by, popular will. It matters little that popular pressures may not lead most directly, or at all, to the truth sought by scientists.

Scientists may ask, rightly, whether the popular will should matter in science. Popular pressure could seriously reduce the long-term viability of the scientific enterprise, while lack of constraint on popular will could lead to widespread and damaging "antiscientific" attitudes. But in a democratic society, citizens must be allowed to choose between supporting science and supporting other valued enterprises. Even though science may be the pursuit of the truth, it is still only one pursuit among many that citizens value. As political philosopher John Rawls (1971) argues, it is the responsibility of democratic governments to pursue justice, which means allowing citizens to pursue their own ideas of what is good rather than directing them to pursue any particular good. Although the idea of public referenda to approve or reject a Grand Unified Theory in physics is ludicrous, there is no escaping public voting on funding research projects like the SSC that could lead to such a theory. The latter votes have all the same risks as the former ones—a good theory may not be found, or a bad one may be promoted, because scientists lack the data that might come from the SSC. But votes on funding also have the legitimacy of the democratic process behind them. Thus, support for science today is notably political, competitive, and demanding. What the populist tension really does is force advocates of scientific research to articulate a compelling public rationale for their preferences and then, like any beneficiary of public expenditures, be held accountable for the outcomes.

The plutocratic tension comes from the possibility that the economic organization that is best for the rapid development of science may be at odds with the economic organization necessary for democracy to flourish. This tension is obvious in concerns about the concentration of R&D funding at a small number of "elite" research institutions, as well as in worries about the real growth of the R&D budget during the 1980s, when most other domestic programs contracted. It is also evident in concern over the growing fuzziness of the boundary between public and private interests, as public employees and private firms are now more able to benefit financially from the fruits of publicly funded research.[23] Another expression of the plutocratic tension is the fear that the benefits of science-based technology—from the profits reaped from new drugs to the conveniences of consumer technologies—most often accrue to those who already possess the largest share of society's bounty. The sense behind the plutocratic tension is that there is a Matthew effect not only within science (Merton 1973 [1968]), but between science and other portions of society: Science is rich and privileged, and because of that wealth and privilege, science becomes richer and more privileged still.

The plutocratic tension exists because democracies require some degree of equality among citizens, usually construed as equal protection under the law and equal opportunity. Democratic equality is usually not taken to mean equality of abilities; however, it is often difficult in practice to distinguish between abilities and opportunities, especially in research, where achievement is a cumulative product of education and equipment as well as intellect. For example, although there is little empirical evidence, it might be that an extreme concentration or clustering of research capital produces the greatest progress toward scientific truth. But it does not follow that such a clustering produces the greatest social, economic, or educational benefits from research. The form of scientific organization that leads to the greatest truth or even to the greatest economic benefits does not necessarily lead to the most just or democratic society.

The exclusionary tension results from the fact that democratic processes and goals are not as compatible with scientific processes and goals as is sometimes thought. The most straightforward example of the exclusionary tension is that the requirements for participation in scientific decision making are higher—for instance, being a scientist or other kind of expert—than for most democratic decision making. Democratic decision making encourages and expands participation; scientific decision making limits it.[24]

Conflicts over the definition of scientific fraud and over how accusations of fraud should be adjudicated provide a clear example of the exclusionary tension. In the early days of the Office of Scientific Integrity (OSI) at NIH, for example, investigations tried to "arrive at the scientific truth as quickly as possible."[25] Investigations focused on "the reliability of data, not the reliability of witnesses."[26] This emphasis on substantive issues ended up slighting procedural issues. OSI claimed that, under this "scientific dialogue method" of investigating misconduct, it did not need to permit the cross-examination of witnesses. As Guston has written about the OSI rules and the exclusionary tension, "If democracy is indeed a process, science as a pursuit of truth may derail it when it deviates from some scientifically defined truth; similarly, democratic processes may reject determination of empirical truth in favor of some conception of fair treatment" (1993a: 49). Or, as Robert Dahl has written more generally about the problem of allowing experts to guard democracy against substantively incorrect decisions, "Carried to an extreme, the insistence that substantive results take precedence over processes becomes a flatly antidemocratic justification for guardianship and 'substantive democracy' becomes a deceptive label for what is in fact a dictatorship" (1989: 163).

The Future of the Contract

This introduction has tried to provide a set of perspectives on the relationship between the federal government and university science. It has not tried to provide prescriptions for writing a new contract or for revising the old one—tasks that have been undertaken by others in both the scientific and the political communities (including some of the contributors to this volume). Our principal conclusion, however, is relevant to thinking about the future of relations between science and politics. It can be stated in a few words: There is no going back to the "good old days" when the social contract for science was first negotiated.

The reason there can be no return rests on the history we have summarized. The image of a time when government provided the money and science provided the results is oversimplified. It applied best to the abnormal years immediately following World War II when the contract was negotiated under the assumption of a near-identity of interest between science and government. But the world of the late 1940s and early 1950s was different from today in almost every way. Moreover, the "good old days" were not always that good. Conflicts—some unique to their time, others replicated more recently—surfaced early in the postwar period.

Even more important are irreversible institutional changes in both science and government. The scientific research community today is vast, expensive, and far more tightly integrated into the economy and social fabric of the nation than ever before. The advent of "imperial" and "managerial" government in both executive and legislative branches gives the presidency and Congress a new interest in, and a new capacity for, detailed oversight of what has become one of the larger discretionary items in the federal budget. Finally, and most important of all, the end of the Cold War, accompanied by a persuasive de-idealization of science and technology, means that older rationales for scientific research have lost much of their power.

One result of these changes, we believe, has been a clearer vision of the essential tension between science and democracy. That tension can be summarized by pointing to the opposition between the values of participation and the pursuit of justice in democratic politics and the values of inquiry and the pursuit of truth in science. The gap between participation and truth was to a large extent papered over in the immediate postwar era, just as it had been almost completely hidden during wartime.[27] But as time went on the tension between science and democracy has emerged with new clarity, and this tension underlies the growing sense of crisis between research universities and the federal government.

The necessary underpinning for any new contract between science and government is a shared understanding that these enterprises are very different in their presumptions and organization. Concretely, this means that scientists as scientists are rarely politically adept or knowledgeable, while politicians are only occasionally qualified to judge the details of scientific claims and disputes.

Yet it also means that scientists and politicians must concede to each other some role in their own enterprises. The scientific community must confront more directly than in the past the fact that it is in competition for federal funding with a large number of other plausible and meritorious projects. Like it or not, if science expects public support, it moves into an arena where it must be political in the best sense—and possibly the worst—in order to justify its claims.[28]

By being "political," we do not simply mean joining the horde of lobbyists who push the competing claims of their clients on Congress and the executive—although in American society lobbying is a time-honored and appropriate activity. We also mean recognizing and responding to the political embeddedness of the scientific community in public attitudes and public policy. The scientific community, and the research universities in which this community is rooted, must undertake an educational role with a double purpose: first, to make clear the nature and workings of science; and second, to bring to the nonscientific community those insights, findings, theories, outlooks, and facts that can indeed contribute to the public good.

In both regards, university science has only begun to explore its role. University scientists need to participate more actively in broadly educational activities such as the training of expert science and technology journalists, along with focused pedagogic activities such as collaboration with educators in primary and secondary schools to improve scientific literacy and the understanding of scientific methods and principles.

Above all, public attitudes toward science and public participation in decision making about scientific priorities must become part of the analysis of science and public policy. The traditional prescription of educating the public in science and technology—to allow their inclusion in the scientists' game—is only a partial solution. The complement to public education must be the acceptance of the public's reasonable concerns, demands, and fears—in other words, educating scientists to play the public's game. Indeed, the traditional resolution of science and public policy into "policy for science" and "science in policy" (Brooks 1968: 85), which excludes the public both grammatically and conceptually, needs to be revised.

Given the inevitable fact that U.S. science competes with other good purposes and institutions for the favorable opinion of the public and the financial support of a democratic government, and given the end of the Cold War, the future relationship of science and government depends heavily on the scientific community's capacity to articulate a plausible rationale for the public support of science. In the past, even scientists adamantly opposed to military research indirectly benefited from the general aura of support for science created by its contributions to the nation's military readiness. Today, as military preparedness yields to international economic competitiveness and domestic wellbeing on the list of national priorities, continuing support for science will depend upon the scientific community's willingness and capacity to contribute to the resolution of economic and domestic problems.

What this requires, we believe, is more vigorous and sustained programs of outreach to the public, public administrators, leaders of the private sector, and lawmakers. If universities do indeed have a contribution of knowledge, perspectives, and understanding to make, it is no longer enough—if it ever was enough—for researchers to wait in their offices for the telephone to ring. More proactive and collaborative projects are required. At MIT, for example, we think of the long-term collaboration between social scientists, led by Suzanne Berger, and leaders of the Joint Chiefs of Staff and the defense community in Seminar XXI, a series of workshops and retreats organized around alternative perspectives on international issues. In the same vein, Eugene Skolnikoff and Claude Canizares are creating a series of workshops on technology-related issues for senior congressional staff members. The Leaders for Manufacturing Program is an experiment in university-industry collaboration designed to produce a new generation of managers trained in the best modern manufacturing methods. At a more general level, the Knight Science Journalism Fellowship Program has contributed to the education of more than one hundred leading science and technology journalists and media experts over the past ten years.

The scientific community must initiate more activities like these, projects that move beyond lobbying to outreach and education, activities that constitute a series of "minicontracts" that draw on the abilities of the scientific community to answer the needs of particular constituencies. It is not enough for the scientific community to claim that their work is "useful"; the relevance of scientific knowledge and perspectives to the public interest must be demonstrated in concrete projects.

Taking such initiatives will be difficult for those scientists whose talents lie in the laboratory rather than in public dialogue and exposition. It is for this reason that the community as a whole needs to encourage and support those gifted teachers and interlocutors whose enthusiasm for science impels them to try to share its relevance and beauty with others. Failure to offer such support could be tantamount to undercutting public support for science.

The changed world of modern science and modern government means that it is imperative to search for and begin to define a new contract, or series of contracts, between the institutions of democracy and the institutions of science. The scientific community needs to reach out to justify its claim on public resources by demonstrating where and how it is relevant in solving public problems. Science needs to earn the confidence of the public and the government, and to enhance its contribution to the general welfare. Conversely, government will need to devise better ways of demarcating the boundaries between what is legitimately political and what is legitimately scientific, of determining the overall levels of support for science and technology and, within that overall budget, of devising rational and fair ways of choosing among distinct and competing scientific and technological priorities.

New contracts are needed, contracts that recognize the increasing interdependence of science and democracy but that also recognize the futility of searching for an end to all tension between them. To be itself, science must continue to pursue scientific truth, and to this end, it will continue to be meritocratic, even elitist. Democratic government, on the other hand, must institute processes that create fairness through egalitarianism and participation. These processes will often deviate from the ends or standards that scientists might prescribe. The attempt to run science on democratic principles would destroy science; but that does not mean that the existing institutions and processes of science are responsive enough to democracy. The attempt to run government on scientific principles would destroy democracy; but that does not mean that our current politics is sufficiently informed or advised by science.

The old contract between science and government was fragile partly because it denied these tensions. It has outlived its usefulness because times have changed since the late 1940s. The new contract as it evolves must seek to comprehend the blurred boundaries between politics and science, while recognizing that the tension between them is intrinsic. Immense benefits can be won, both for democracy and for science, if their interactions can be managed fairly, intelligently, and with mutual respect.

Notes

1. What these reports do have in common is that they each stress the necessity of, in the words of the Carnegie Commission (1992), "linking science and technology to societal goals" without destroying the productivity of the current system.

2. The Mansfield Amendment of 1970 forbade the Department of Defense from funding research that was unrelated to military applications. The amendment passed that year with the support of a coalition of the ends: liberals who wanted to reduce military spending and conservatives who wanted to reduce "useless" federal spending. The next year, however, the amendment was repealed, but a "Ghost Mansfield Amendment" continued to hang over military research, threatening the funding of basic research should it appear irrelevant (Bromley 1988: 20).

3. The federal government funded about 61 percent of all basic research in 1991, but conducted only about 12 percent. Similarly, the federal government funded about 35 percent of all applied research that year, but conducted again about 12 percent (NSB 1991: 93).

4. In a 1990 court case in which a federally funded researcher's property interest in his grant was at issue (for due process rights), the judge commented that "The grant awards made to the Board of Regents are not made for [the researcher's] benefit, but for the benefit of the public that may enjoy the fruits of his research." Legally, grants are discretionary actions of the agencies involved. See *Abbs v. Sullivan*, U.S. District Court for the Western District of Wisconsin, 90-C-470-C, December 20, 1990.

5. See Chubin and Hackett (1990) for the most comprehensive treatment of peer review, its applications, and its faults.

6. Generally, critics maintain that the social contract model cannot sustain professional norms for two reasons. First, it assumes a prior norm that contracts should be adhered to; and second, it fails to address conduct toward persons who are not parties to the contract, e.g., clients (Bayles 1983; Schmaus 1983).

7. For discussions of auditing scientific data, see Pincus (1990). See also J. Mervis, "Random audits of raw data?" *The Scientist* (November 28, 1988):1; and B. Davis, "Fraud: Why auditing laboratory records is a bad idea," *The Scientist* (February 20, 1989):9.

8. Note the generational aspect of this social contract for scientists.

9. There are few systematic histories of postwar science and technology policy. Smith (1990) concentrates mostly on funding patterns and shifting government roles in basic research and technology development. U.S. Congress (1986) touches on several enduring trends in science policy such as the tension between the centralization of the research effort and pluralistic funding sources. Other histories concentrate on particular disciplines (e.g., Kevles 1978) or particular funding agencies (e.g., England 1982; Harden 1986; Sapolsky 1990).

10. NSF created the RANN program in 1971 by expanding what had been the Interdisciplinary Research on Problems Relevant to Our Society (IRPOS) program. RANN funding constituted 8.6 percent of all NSF funding in 1972 and rose to 13 percent in 1974, but fell to 10 percent in 1976 (Smith and Karlesky 1977: 34). The program was eliminated in 1978. Like RANN, the War on Cancer at NIH was a program that followed on the heels of political criticism. The War on Cancer produced a slight shift in emphasis at NIH that would favor contracts over grants and clinical research over basic research (Rettig 1977). The Mansfield Amendment is discussed in note 3 above.

11. Based on recommendation NSF01 of the National Performance Review, President Clinton issued Executive Order 12881 on 23 November 1993 to establish NSTC, which is to be parallel to the National Security Council and the National Economic Council, and coordinate science and technology policy in the executive branch.

12. For changing views of the presidency, see Moe (1993). For the increased roles of science advisors and advisory committees, see Jasanoff (1990) and Smith (1992).

13. For the general "resurgence" of Congress, see Sundquist (1981). Smith (1985) and Shepsle (1989) document new forms of decision making in Congress that are less centralized and directed. Huntington (1965) and Aberbach (1990) both describe the increase of congressional oversight over executive agencies. For summaries of the roles of the legislative agencies, see Dodd and Schott (1986), but for OTA see Bimber (1992) and for GAO see Mosher (1979), Walker (1986), and Trask (1991). Light (1993) describes the use of Inspectors General in the "search for accountability" in government.

14. These numbers are scientists and engineers engaged in R&D. The source for the 1965 and 1988 data is NSB (1991: 301). Earlier data are not comparable.

15. Dresch and Janson (1987, 1990) present an econometric model of the relationship between the recruitment of scientists and the output of scientific papers. They find that a point of diminishing returns exists such that the marginal addition of scientists performing basic research adds little to the

literature and that such marginal scientists would produce more benefits if they turned instead to careers in technology or education. Extending the analogy with social welfare programs, the technology transfer legislation of the 1980s could also be viewed as a "welfare to workfare" program in which scientists are encouraged to patent research products for additional money for their laboratories and to make contacts with private firms who could help support future research.

16. See Cohen and Noll (1991) for a discussion of "technology pork barrel" programs like the space station.

17. See M. Wines, "House kills the supercollider, and now it might stay dead," *New York Times* (October 20, 1993): A1, A20.

18. Killian (1982) provides a fine rendering of this period.

19. "Technics-out-of-control" has been a theme in both in American political thought (Winner 1977) and in popular culture. The two most popular movies of the summer of 1993, for example, involved themes of the evils of commercialized technology. See Gould (1993). See also G. Kolata, "Forget the butler; the medical industry did it," *New York Times* (October 17, 1993): E2.

20. This title is taken from Guston (1993a,b), which, along with Graham (1993) and Laird (1993a) is the source for some of the discussion below.

21. Representative George Brown organized the Task Force on the Health of Research in May 1991 (102nd Congress) to "investigate the widespread reports of 'stress' in the federal research system, and to define a more active role for the Committee in the forging of science policy for the national good" (U.S. Congress 1992: v). The Science Policy Task Force was an ongoing inquiry directed by Representative Don Fuqua in 1985 and 1986 that held some twenty-four hearings and produced an additional ten reports on all aspects of science policy. The Fountain Committee was the informal name of the House Government Operations Committee that, under the chairmanship of Representative Lawrence Fountain, investigated many aspects of the research program at NIH in the 1960s (Greenberg 1967; Strickland 1972). The Elliott Committee, named for Representative Carl Y. Elliott (D-AL), was a Select Committee on Government Research created in 1963 that held a number of hearings and produced ten reports during the 88th Congress (Kevles 1978; U.S. Congress 1980, 1986). The Daddario Subcommittee, named for its chairman Emilio Daddario (D-CT), was created by the House Science and Astronautics Committee in 1963, in part to protect its jurisdictional boundaries in response to the creation of the Elliott Committee (U.S. Congress 1980; U.S. Congress 1986). The Joint Committee on Atomic Energy in the 1950s was the first and perhaps the most successful standing congressional committee with an explicit jurisdiction over a large portion of the research establishment (Green and Rosenthal 1963). The Kilgore Committee, with Senator Harley Kilgore (D-WV) as its chairman, performed substantial work at the end of World War II to promote the continuation of federally sponsored research, leading to the creation of NSF (England 1982).

22. For more on the Allison Commission, see Dupree (1957: ch. 11); Kevles (1978: 51–59); Manning (1988); and Guston (1994).

23. Sociologist Robert K. Merton (1973 [1942]) recognized some form of this tension when he observed that the scientific norm of the communal ownership of research results was incompatible with the economic system of incentives in modern, capitalist democracies.

24. Laird (1993a, b) proposes a system of "participatory analysis" to reduce the tensions between democratic participation and technical decision making.

25. Quoted in Guston (1993b: 48). OSI is now known as the Office of Research Integrity (OTI).

26. Quoted in Guston (1993b: 48).

27. French sociologist Bruno Latour (1991) suggests that the conceptual separation of politics and science was an essential part of the modern world, and he recommends that by rejecting the separation, society may be more able to manage complex interdisciplinary problems like crime and pollution.

28. Gerald Holton (1993) has noted that one element in the current political attacks on science has analogies to the critique of science by nineteenth- and twentieth-century Romantic philosophers, which was adopted by Nazi ideologists, among others. See also Berlin (1992: esp. ch. 7).

Bibliography

Aberbach, J. 1990. *Keeping a Watchful Eye: The Politics of Congressional Oversight.* Washington, DC: The Brookings Institution.

Alic, J. A., Branscomb, L. M., Brooks, H., Carter, A. B. and Epstein, G. L. 1992. *Beyond Spinoff: Military and Commercial Technologies in a Changing World.* Boston: Harvard Business School Press.

Allen, J., ed. 1970. *March 4: Scientists, Students, and Society.* Cambridge: MIT Press.

American Association for the Advancement of Science (AAAS). 1994. *AAAS Report XIX: Research and Development, FY 1995.* Intersociety Working Group. Washington, DC: AAAS Press.

Barke, R. P. 1990. Beyond the "Endless Frontier": Changes in the political context of American science and technology policy. Paper presented at the annual meeting of the Society for Social Studies of Science, Minneapolis, October.

Bayles, M. D. 1983. *Professional Ethics.* Belmont, CA: Wadsworth Press.

Berlin, I. 1992. *The Crooked Timber of Humanity.* New York: Random House.

Bimber, B. 1992. *Institutions and Information: The Politics of Expertise in Congress.* Unpublished Ph.D. dissertation, Department of Political Science. Cambridge: MIT.

Branscomb, L. M., ed. 1993. *Empowering Technology: Implementing a U.S. Strategy.* Cambridge: MIT Press.

Bromley, D. A. 1988. Science advice: Past and present. In K. Thompson, ed., *The Presidency and Science Advising,* vol. 5, 1–25. New York: University Press of America.

Bromley, D. A. 1993. National science and technology policy. Presented as the NIH Director's Lecture, Bethesda, MD, January.

Brooks, H. 1968. *The Government of Science.* Cambridge: MIT Press.

Brooks, H. 1990. Can science survive in the modern age? *National Forum* 71(4):31–33.

Brooks, H. 1993. Research universities and the social contract for science. In L. M. Branscomb, ed., *Empowering Technology: Implementing a U.S. Strategy,* 202–234. Cambridge: MIT Press.

Brown, G. E., Jr. 1991. A perspective on the federal role in science and technology. In M. O. Meredith, S. D. Nelson, and A. H. Teich, eds., *AAAS Science and Technology Policy Yearbook, 1991,* 23–33. Washington, DC: AAAS Press.

Brown, G. E., Jr. 1992. The objectivity crisis. *American Journal of Physics* 60:779–781.

Bruce, R. V. 1987. *The Launching of American Science, 1846–1876.* Ithaca: Cornell University Press.

Bush, V. 1990 [1945]. *Science: The Endless Frontier.* Washington, DC: National Science Foundation.

Carnegie Commission on Science, Technology, and Government. 1992. *Enabling the Future: Linking Science and Technology to Societal Goals.* New York: Carnegie Corporation of New York.

Chubin, D. E., and Hackett, E. J. 1990. *Peerless Science: Peer Review and U.S. Science Policy.* Albany: State University of New York.

Cohen, L. R., and Noll, R. G., eds. 1991. *The Technology Pork Barrel.* Washington, DC: The Brookings Institution.

Congressional Budget Office (CBO). 1993. A review of Edwin Mansfield's estimate of the rate of return from academic research and its relevance to the Federal budget process. Staff memorandum.

Council of Biology Editors (CBE). 1983. *CBE Style Manual: A Guide for Authors, Editors, and Publishers in the Biological Sciences.* 5th ed. Bethesda, MD: CBE.

Dahl, R. A. 1989. *Democracy and its Critics.* New Haven: Yale University Press.

Dertouzos, M., Lester, R., and Solow, R. 1989. *Made in America: Regaining the Productive Edge.* Cambridge: MIT Press.

Dodd, L. C., and Schott, R. L. 1986. *Congress and the Administrative State.* New York: Macmillan.

Dresch, S. P., and Janson, K. R. 1987. Giants, pygmies, and the social costs of fundamental research, or Price revisited. *Technological Forecasting and Social Change* 32:323–340.

Dresch, S. P., and Janson, K. R. 1990. Recruitment and accomplishment in fundamental science: A generalization of the "giants, pygmies" model. *Technological Forecasting and Social Change* 37:39–58.

Dupree, A. H. 1957. *Science in the Federal Government: A History of Policy and Activities to 1940.* Cambridge: Harvard University Press.

Edsall, J. T. 1988 [1955]. Government and the freedom of science. In R. Chalk, ed., *Science, Technology, and Society: Emerging Relationships*, 24–27. Washington, DC: AAAS Press.

Eisenhower, D. D. 1961. Farewell radio and television address to the American people. In *Public Papers of the Presidents of the United States, 1961*, 1035–1040. Washington, DC: U.S. Government Printing Office.

England, J. M. 1982. *A Patron for Pure Science.* Washington, DC: National Science Foundation.

Ezrahi, Y. 1990. *The Descent of Icarus: Science and the Transformation of Contemporary Democracy.* Cambridge: Harvard University Press.

Foerstal, H. N. 1993. *Secret Science: Federal Control of American Science and Technology.* Westport, CT: Praeger.

Government-University-Industry Research Roundtable (GUIRR). 1992. *Fateful Choices: The Future of the U.S. Academic Research Enterprise.* Washington, DC: National Academy Press.

Gould, S. J. 1993. Dinomania. *New York Review of Books* 40(12 August):51–56.

Graham, G. 1993. The necessity of the tension. *Social Epistemology* 7(1):25–34.

Green, H., and Rosenthal, A. 1963. *Government of the Atom.* New York: Atherton Press.

Greenberg, D. S. 1967. *The Politics of Pure Science.* New York: New American Library.

Guston, D. H. 1993a. Resolving the tension in Graham and Laird. *Social Epistemology* 7(1):47–60.

Guston, D. H. 1993b. The essential tension in science and democracy. *Social Epistemology* 7(1):3–23.

Guston, D. H. 1993c. *The Social Contract for Science: Congress, the National Institutes of Health, and the Boundary between Politics and Science.* Unpublished Ph.D. dissertation, Department of Political Science. Cambridge: MIT.

Guston, D. H. 1994. Congressmen and scientists in the making of science policy. *Minerva* 32(1):25–52.

Harden, V. A. 1986. *Inventing the NIH: Federal Biomedical Research Policy, 1887–1937.* Baltimore: Johns Hopkins University Press.

Holton, G. 1993. The value of science at the "end of the modern era. Paper presented at the Sigma Xi Forum on Ethics, Values, and the Promise of Science, Chapel Hill, February.

Huntington, S. P. 1965. Congressional responses to the twentieth century. In D. B. Truman, ed., *The Congress and America's Future*, 5–31. Englewood Cliffs, NJ: Prentice-Hall.

Jasanoff, S. 1990. *The Fifth Branch: Science Advisors as Policymakers*. Cambridge: Harvard University Press.

Kevles, D. J. 1978. *The Physicists: The History of a Scientific Community in Modern America*. New York: Knopf.

Kevles, D. J. 1992. Foundations, universities, and trends in support for the physical and biological sciences, 1900–1992. *Daedalus* 121(4):195–236.

Killian, J. R. 1982. *Sputnik, Scientists, and Eisenhower*. Cambridge: MIT Press.

Laird, F. 1993a. Participating in the tension. *Social Epistemology* 7(1):35–46.

Laird, F. 1993b. Participatory analysis, democracy, and technological decisionmaking. *Science, Technology, and Human Values* 18:341–361.

Latour, B. 1991. The impact of science studies on political philosophy. *Science, Technology, and Human Values* 16:3–19.

Lederman, L. M. 1991. *Science: The End of the Frontier?* Washington, DC: AAAS Press.

Light, P. C. 1993. *Monitoring Government: Inspectors General and the Search for Accountability*. Washington, DC: The Brookings Institution.

Manning, T. G. 1988. *U.S. Coast Survey vs. Naval Hydrographic Office: A 19th-Century Rivalry in Science and Politics*. Tuscaloosa: University of Alabama Press.

Mansfield, E. 1991. Academic research and industrial innovation. *Research Policy* 20:1–12.

Merton, R. K. 1973 [1968]. The Matthew Effect in science. In N. W. Storer, ed., *The Sociology of Science*, 439–459. Chicago: University of Chicago Press.

Merton, R. K. 1973 [1942]. The normative structure of science. In N. W. Storer, ed., *The Sociology of Science*, 267–278. Chicago: University of Chicago Press.

Moe, T. 1993. Presidents, institutions, and theory. In G. C. Edwards III, J. H. Kessel, and B. A. Rockman, eds., *Researching the Presidency: Vital Questions, New Approaches*, 337–385. Pittsburgh: University of Pittsburgh Press.

Mosher, F. C. 1979. *The GAO: The Quest for Accountability in American Government*. Boulder, CO: Westview Press.

Mukerji, C. 1989. *A Fragile Power: Scientists and the State*. Princeton: Princeton University Press.

National Science Board (NSB). 1973. *Science Indicators, 1972*. Washington, DC: U.S. Government Printing Office.

National Science Board (NSB). 1991. *Science and Engineering Indicators, 1991*. 10th ed. NSB-91-1. Washington, DC: National Science Foundation.

National Science Board (NSB). 1992. *A Foundation for the 21st Century*. Washington, DC: National Science Foundation.

Nelkin, D. 1972. *The University and Military Research: Moral Politics at MIT.* Ithaca: Cornell University Press.

Office of Technology Assessment (OTA). 1991. *Federally Funded Research: Decisions for a Decade.* OTA-SET-490. Washington, DC: U.S. Government Printing Office.

Perrow, C. 1984. *Normal Accidents: Living with High-Risk Technologies.* New York: Basic Books.

Pincus, K. V. 1990. Financial auditing and fraud detection: Implications for scientific data audit. *Accountability in Research: Politics and Quality Assurance* 1(1):53–70.

Polanyi, M. 1962. The republic of science: Its political and economic theory. *Minerva* 1:54–73.

Polanyi, M. 1964 [1946]. *Science, Faith and Society.* Chicago: University of Chicago Press.

Press, F. 1988. The dilemma of the golden age. Address to members, 125th annual meeting of the National Academy of Sciences, Washington, DC.

Price, D. de S. 1986. *Little Science, Big Science . . . and Beyond,* with forewords by R. K. Merton and E. Garfield. New York: Columbia University Press.

Price, D. K. 1954. *Government and Science: Their Dynamic Relation in American Democracy.* New York: New York University Press.

Primack, J., and von Hippel, F. 1974. *Advice and Dissent: Scientists in the Political Arena.* New York: Basic Books.

Rawls, J. 1971. *A Theory of Justice.* Cambridge: Harvard University Press.

Rettig, R. 1977. *Cancer Crusade: The Story of the National Cancer Act of 1971.* Princeton: Princeton University Press.

Rottenberg, S. 1968. The warrants for basic research. In E. Shils, ed., *Criteria for Scientific Development: Public Policy and National Goals,* 134–142. Cambridge: MIT Press.

Roush, W. 1993. Learning from technological disasters. *Technology Review* 96(6):50–58.

Sapolsky, H. 1990. *Science and the Navy: The History of the Office of Naval Research.* Princeton: Princeton University Press.

Schmaus, W. 1983. Fraud and the norms of science. *Science, Technology, and Human Values* 8:12–22.

Shepsle, K. A. 1989. The changing textbook Congress. In J. E. Chubb and P. E. Peterson, eds., *Can the Government Govern?* Washington, DC: The Brookings Institution.

Smith, B. L. R. 1990. *American Science Policy Since World War II.* Washington, DC: The Brookings Institution.

Smith, B. L. R. 1992. *The Advisors: Scientists in the Policy Process.* Washington, DC: The Brookings Institution.

Smith, B. L. R., and Karlesky, J. J. 1977. *The State of Academic Science: The Universities in the Nation's Research Effort.* New York: Change Magazine Press.

Smith, S. S. 1985. New patterns of decisionmaking in Congress. In J. E. Chubb and P. E. Peterson, eds., *The New Direction in American Politics,* 203–234. Washington, DC: The Brookings Institution.

Steelman, J. R. 1947. *Science and Public Policy.* The President's Scientific Research Board. Washington, DC: U.S. Government Printing Office.

Strickland, S. P. 1972. *Politics, Science and Dread Disease: A Short History of United States Medical Research Policy.* Cambridge: Harvard University Press.

Sundquist, J. L. 1981. *The Decline and Resurgence of Congress.* Washington, DC: The Brookings Institution.

Taubes, G. 1993. *Bad Science: The Short Life and Bad Times of Cold Fusion.* New York: Random House.

Trask, R. R. 1991. *GAO History, 1921–1991.* GAO/OP-3-HP. Washington, DC: GAO.

Turner, S. P. 1990. Forms of patronage. In S. E. Cozzens and T. F. Gieryn, eds., *Theories of Science in Society,* 185–211. Bloomington: Indiana University Press.

U.S. Congress. 1980. *Toward the Endless Frontier: History of the Committee on Science and Technology, 1959–1979.* House of Representatives, Committee Print. Washington, DC: U.S. Government Printing Office.

U.S. Congress. 1986. *A History of Science Policy in the United States, 1940–1985.* Science Policy Study Background Report No. 1. Task Force on Science Policy, House of Representatives, Committee on Science and Technology, 99th Cong., 2nd sess. Washington, DC: U.S. Government Printing Office.

U.S. Congress. 1992. *Chairman's Report to the Committee on Science, Space, and Technology.* Task Force on the Health of Research, House of Representatives, Committee on Science, Space, and Technology, 102nd Cong., 2nd sess. Washington, DC: U.S. Government Printing Office.

Walker, W. E. 1986. *Changing Organizational Culture: Strategy, Structure, and Professionalism in the U.S. General Accounting Office.* Knoxville: University of Tennessee Press.

Wang, J. 1992. Science, security, and the Cold War. *Isis* 83:238–269.

Weiner, C. 1970. Physics in the Great Depression. *Physics Today* 23:31–37.

Winner, L. 1977. *Autonomous Technology: Technics-Out-of-Control as a Theme in Political Thought.* Cambridge: MIT Press.

Winner, L. 1986. *The Whale and the Reactor: A Search for Limits in an Age of High Technology.* Chicago: University of Chicago Press.

Zuckerman, H. 1977. Deviant behavior and social control in science. In E. Sagarin, ed., *Deviance and Social Change,* 87–138. Beverly Hills: Sage Publications.

Zuckerman, H. 1984. Norms and deviant behavior in science. *Science, Technology, and Human Values* 9:7–13.

2

Universities, the Public, and the Government: The State of the Partnership

Charles M. Vest

In the post–World War II era, universities have been the intellectual centerpiece of federal policy for research, education, and innovation. Federal research and development grants and contracts were to fund new discoveries, train new researchers, and provide a steady stream of ideas to improve security, health, and industry. The government's expectations of universities were explicit, but the policies, mechanisms, and measurements to realize and document them were left largely implicit.

The "victory" of the United States and its allies in the Cold War has encouraged a rosy assessment of the contribution of science and technology to the achievement of military superiority. However, social, economic, and environmental difficulties have encouraged a negative assessment of the contribution of science and technology to civil life. This assessment, and the consequent emphasis on accountability based on explicit policies, mechanisms, and measurements, has added to the perceived burdens of the research universities.

In this chapter, Charles M. Vest, president of the Massachusetts Institute of Technology (MIT), considers how the relationship between the federal government and the research universities has changed, and what might be done to correct for some of the more problematic changes. President Vest is well-placed for this discussion, for MIT in the early 1990s has been a protagonist in high-profile conflicts with the federal government on issues such as scientific integrity, the billing of indirect research costs, technology transfer, and the administration of financial aid.

Vest lauds research universities for their achievements and contributions, but he also meets the concerns of critics with candor. He believes that research universities are a "bargain" for society's investment in human capital, but he also believes that universities must take on themselves the responsibility of educating the public about the value of this investment as well as the role of universities in making it pay off.

Beyond education, Vest makes recommendations in four areas of the univeristy-government partnership: funding, management, distribution of resources, and academic integrity. His recommendations can all be characterized as compromises, asking both partners to make concessions in order to forge a stronger relationship in the future. Vest focuses on academic integrity, the moral and intellectual foundation of any partnership. Although challenges to academic integrity arise both within and outside of the research enterprise, Vest argues that academic misconduct is rare and can be made rarer still if universities and the individuals within them take seriously their duties to educate students in the virtues of integrity and "ethical rigor" as well as methodology and intellectual rigor. This relationship between truth and integrity that Vest identifies as the core of the duty of the research university also lies at the most fragile passage of the contract between science and society. [Eds.]

Introduction

The past several years have seen a troubling erosion of the strong partnership between research universities in the United States and the federal government; and it remains to be seen whether this erosion is just around the edges and short-lived, or central and long-lived. Even more troubling is the resonance that various criticisms of academia are finding in popular opinion. This trouble comes at a time when it is absolutely essential that the nation invest heavily in the development of its human capital, that is, in its people and their ideas.

For five decades, the fundamental driving forces for federal support of university research and education have been national security and health care. In the post–Cold War world, it is entirely possible that military security will no longer dominate congressional priorities, and therefore the rationale for university support may drastically change or diminish (see Sapolsky, this volume). Before addressing some of the specific issues that I have dealt with as president of MIT, I would like to suggest one fundamental question of science policy to keep in mind: ten years from now, how should the federal government fund university research and education, and why?

These are exciting and tough times to be in academia. Universities face continuing debates over what should constitute the core of an undergraduate educational experience, tensions stemming from a growing diversity and changing expectations among students and faculty, and inquiries into how they conduct research and charge the sponsors of research for its costs.

One of greatest challenges facing the universities is to listen carefully to what the critics say, to understand if some of these criticisms are valid. To the extent that they are, then universities should address the problems. On the other hand, universities should take care not to behave in a reactionary manner. Rather, universities should remind themselves that they are a powerful and positive force in the larger society. Universities should define themselves and their relations with other social institutions in such a way that they can carry the best values of service and leadership into the future. Universities, particularly research universities, must be self-conscious about the state of their partnership with the government—the representatives of the public and the society they serve.

The Research University

For decades, the research university has served this nation well. Indeed, from virtually any perspective, it has paid enormous dividends in return for the public's trust and investment: dividends in the form of educated leaders in academia, business, and government; dividends in the form of advances in understanding and amelioration of critical problems relating to health, the economy, housing, and the environment; dividends in the form of national security and of new and revitalized industries that have spun off from university research; dividends in the form of increased understanding of our physical, social, and natural worlds.

These dividends, however, are often unrecognized or undervalued by the public. In the past, Americans believed that higher education was singularly important for the betterment of their children's lives, and they were willing to invest public, private, and personal funds to create, sustain, and enhance our public and private universities. And the government looked to the universities of this country to be the centers of the nation's research enterprise.

The resulting system of private and public research universities is unique to the United States. The system blends graduate education, undergraduate education, and research to create unparalleled opportunities for learning and discovery. The system quickly involves young faculty members as full partners in the academic enterprise, helping to keep both education and research forward-looking and robust. The system motivates strong, creative individuals to teach the living essence and the future of their subjects, not just their histories.

One of this system's greatest strengths is that it is not homogeneous. The variety of private and public universities in this country allows for—indeed demands—experimentation, variation, cooperation, and even competition. The resulting synergy is the yeast that keeps the system strong.

But if this system is so good, why are universities losing some of the general respect and support they have enjoyed in the past? The simple answer is that the environment for research universities in the United States has changed, and these changes must be recognized and adapted to, to make for a healthier climate for us all.

The Changing Environment: Five Perspectives

In large measure what has enabled the system of research universities to develop and flourish has been their partnership with the federal government. The federal government has been by far the largest source of funding for research programs and has traditionally been a major source of financial aid for both graduate and undergraduate students. Today, however, there are strains in the system, many of which are manifestations of tensions between the federal government and the research universities.

There are, of course, different assessments of the degree to which the partnership between the federal government and the research universities has eroded. Robert Rosenzweig (1990), then president of the Association of American Universities, has gone so far as to say that there is no partnership, and that it is naive to assume that there could be one within our system of government.

I believe that over time there has been a sense of partnership and that we must reinvigorate it. Surely there is a partnership implicit in the post–World War II decision to have universities serve as the nation's infrastructure for basic research. The crisis of World War II, and even the challenge of Sputnik, led to well-defined goals for federal funding of research universities—funding relating to education and educational opportunity, as well as research.

Today the nation is also in crisis, but the threat and its solution lack clarity. The meeting of minds necessary for the "partnership" may be lacking. Universities and the government seem to be drawing apart from each other in their respective views of the mission and value of the research universities. In order to perceive more clearly the changes in the environment that are occurring, I would like to examine the current situation from several different, and intentionally overstated, perspectives.

The Federal Government

It is clear that universities and the research community are increasingly viewed by many in the government as just another supplicant at the public trough. And they have been successful supplicants: federal funding of academic science and engineering research and development has roughly doubled over the past decade.

Still, the government is besieged by complaints from the university community that support is far too low and that a severe shortfall of Ph.D. scientists and engineers looms just ahead. The scientific community is thought to lack a coherent view of what the national policy for research should be. In particular, the scientific community is seen as being deeply divided over the relative merits of large-scale and single-investigator research (see Kleppner, this volume). And while scientists may argue that universities, and university research in particular, are the keys to a healthy, competitive economy, there is a growing chorus that universities are selling the fruits of federally sponsored research to global competitors for a pittance (see Skolnikoff, this volume). Finally, as some in government look at allegations of dishonesty in the conduct of research, possible conflicts of interest on the part of faculty involved with private business ventures, and charges regarding the indirect costs of research, they wonder whether universities are, in fact, upholding the public trust.

The Public

The general public voices some of the same concerns as the government (see Nelkin, this volume). I suspect, however, that on the whole their attention is caught primarily by the cost and quality of undergraduate education. Over the past decade, as tuition levels have risen faster than inflation, the public has been besieged by books and reports attacking the quality of undergraduate education at the nation's universities. Many of these reports paint a picture of undergraduate teaching as relegated by research-oriented faculty to graduate teaching assistants, whose communication skills are minimal and whose sole interest is their own research. The clear suspicion is that tuition levels are growing rapidly in order to pay for research, which is viewed as something quite apart from education. I suspect that the proverbial person in the street believes that university faculty members lead leisurely, somewhat self-indulgent lives, earn high salaries, and teach very little. To top things off, the public is further subjected to overblown rhetoric from both ends of the political spectrum about "political correctness."

The Faculty

Faculty clearly sense a changing atmosphere within research universities. Older faculty look back with a certain nostalgia on an era in which they could spend more time on research and less on research administration, a time when the leading research universities were better able to fund innovative projects and take the kinds of chances that can lead to sea changes in intellectual endeavors. Younger faculty are fixated on what they view as the mechanics of attaining tenure. They underrate the importance that most of their colleagues and administrators place on their teaching contributions when judging them for tenure. They feel pressed to establish research programs, laboratories, funding, and lengthy publication records, and to do so rapidly in a world where the competition is fierce for grants that continue to shrink in size and duration (see Chubin, this volume).

These younger faculty members believe that research sponsors and university administrators require them to spend most of their time generating entropy rather than useful work. And they wonder if they will ever have time to spend with their families (Czujko, Kleppner, and Rice 1990; Lederman 1991).

Students

Students appear to be satisfied in many dimensions, but seem to have the sense that somehow things should be better still. Undergraduates often believe that they have too many teaching assistants, not enough direct contact with faculty members, and insufficient academic and career counseling. Some feel alienated and do not detect an integrated structure to their campus life. They worry about the accumulation of large debts during their college years.

Graduate students generally seem to find instruction to be very good. But whereas in earlier times graduate students were often well-funded through federal fellowship programs, they are now plagued with concerns about how the vagaries of their mentors' research funding will affect their intellectual development and the quality of their lives. They know that there is a mismatch between the production of doctorates and the near-term job market. And they see the faculty life to which they may aspire as complex, frenetic, and increasingly unattractive.

Administrators

Finally, the senior academic administrators are deeply concerned about the effects on their universities of the budget squeeze, the regulatory spiral, and the politicization of university-government relations. These problems affect both public and private universities, although their severity is generally greater at the latter.

Administrators see the costs of doing what universities expect of themselves, and what society expects of them, growing much faster than their revenues. There is little obvious cause to hope for improvement in any of the major sources of income: tuition, private donations, and federal and state support. Compounding the gloom are the spiraling costs of government regulation, litigation, health care, and scientific equipment, and the steady shift of the financial aid burden to internal funding sources. At MIT, for example, the federal share of funding for scholarships has dropped from 38 percent to 12 percent during the last decade, and the expenditure of the university's unrestricted funds for scholarships has increased by a factor of over six during this time. Other detrimental shifts include changing tax policy, such as the cap on tax-exempt bonding by private universities and the reduction of tax deductibility of appreciated, donated assets. As Donald Kennedy (1990) has noted, this action on tax policy signals a profound change in federal attitude and the onset of a view of universities as merely another interest group.

These are not the only problems. Consider, for example, the growing requirements for cost-sharing in order to win grants and contracts; the escalating salary competition for faculty; and the creation of a new, major, budgetary component to support information technology for teaching, research, and administration. Beyond that, administrators are frustrated by our inability to make major progress in enrolling students and appointing faculty from underrepresented groups, particularly in science and engineering. And as noted above, administrators are alarmed by the accelerating intrusion of politics into the research enterprise, dramatized by the increasing tendency to earmark research funds and to avoid peer review as a normal way of doing business (see Chubin, this volume).

The Research University: Savior or Supplicant?

What is the reality of these views and what does it say about the relationship between the federal government and our research universities?

Research universities are indeed highly dependent on federal funding. As the cost of meeting their needs—to say nothing of their aspirations—continues to outpace the available resources, universities turn to the government with ever greater vigor to help restore the balance. Indeed, in the past decade, research universities have been the beneficiaries of some of the most rapidly growing components of the federal budget.

Universities are supplicants in a narrow sense, but they are also much more. The research universities are the primary means by which the government and the private sector can invest in the future. They are, after all, the prime developers of people and ideas. Universities are not only a good investment, but a bargain at that. They conduct essential research with overhead rates far below those charged by any alternative organization. More important, in the process, they educate the next generation of scholars, researchers, entrepreneurs, public servants, and leaders.

The performance of research universities during the past five decades has, by any measure, been truly exceptional. Yet the public must be reminded, for example, about the industries and jobs that have been generated by university research. MIT graduates, faculty, and staff have founded businesses, mostly technology-based, that have created some 300,000 jobs in Massachusetts and 150,000 jobs in California alone. University researchers developed the first digital computer and continue to play leading roles in the information technology revolution. The recombinant DNA techniques that spawned the modern biotechnology industry were born in university laboratories.

If the public and the government truly understood the value of research universities, I believe they would be stand together with them more firmly as partners. Universities may bemoan a lack of public understanding, but they cannot lay blame. Universities need to make their case far better than they do. They need to be educators. The public and the government need a better understanding of the nature and mission of research universities and the nature of faculty life and accomplishments. They need to know the reasons for the growing costs of education. They need to know, without hyperbole, about the successes of our university research and educational systems over the years—in order to appreciate more fully the role of research universities in developing the nation's potential.

What Is To Be Done?

For the sake of argument, assume that a partnership between the research universities and the federal government does exist. What can be done to reinvigorate this nation's investment in its future? How can the partnership be restructured to provide the best environment for faculty to conduct first-rate research and provide first-rate education? I would like to make a few suggestions. They fall into four general categories: funding, management, distribution of resources, and academic integrity and the values of science.

Funding

What can be done about the imbalance between the needs for federal funding on the part of universities and the scarcity of resources available for these purposes? Left unchecked, this situation will worsen, as entitlements grow voraciously, and as the federal budget is used to care for other pressing domestic problems and to maintain our responsibilities abroad. I would argue that it is in the long-range interests of the nation for the federal government to reorder its priorities, so that it can invest more heavily in human capital, in basic research, and in national industrial competitiveness.

Although I do not believe that the research and education enterprises in this country should be centrally managed, I do believe that the nation and the scientific community should assess their goals and priorities, gain public consensus for them, and develop federal research and education budgets that, in some measure, reasonably reflect the costs of attaining these goals for the development of human capital and the enhancement of basic and applied research.

Management

As part of this effort, I believe that the federal government should take major strides to reduce regulatory and bureaucratic requirements imposed upon universities, following an analysis of costs and benefits. Implementing the 1986 recommendations of a study panel of the Office of Science and Technology Policy (OSTP 1986, also known as the Packard-Bromley report) would be a good start (see Likins and Teich, this volume). But in order to ask credibly for modified approaches from federal sponsors, universities must simultaneously take steps to be as financially and administratively efficient as possible. Universities could take a page from the book of lessons learned by much of U.S. industry during the past decade. Total quality management, continuous quality improvement, and lean production may well have analogs in university research and education that could be pursued. Universities must focus their mission, pruning as well as fertilizing their enterprise, in order to keep it robust.

Distribution of Resources

Third, universities must address the difficult issue of how to optimize the use of federal funding for research and graduate education, and to do so in a manner that continues to allow for local initiative, innovation, and entrepreneurship. In my view, there are a number of steps that should be taken:

- Greatly increase the number of federally funded fellowships and cover the real educational costs associated with them. (This will help strengthen the free market in graduate education and shorten the time to degree, which is increasing at a troublesome rate.)
- Foster the diversity of kinds of universities in the country, reversing the growing trend toward homogeneity and recognizing the value of different emphases of mission in different schools.
- Move to covering more nearly the real cost of federally sponsored research and provide for better continuity of support. These are crucial changes for private universities if they are simultaneously to restrain tuitions, bear a major share of student financial aid, and maintain their physical plants and infrastructures.
- Return to strong dominance of the peer review system by investing in high quality wherever it is found. Basic fairness will, in the long run, come about by ensuring access to graduate education and research for bright students from all geographic regions, races, economic backgrounds, and both genders, even if research funding ends up somewhat more focused on a modest number of public and private institutions across the country.
- Finally, universities must recommit themselves to the highest standards in the conduct of research and the stewardship of the

public trust. No suggestions regarding a renewed partnership will go far unless universities are able to address forthrightly the issue of academic integrity. If universities are to warrant the public trust—if they are to be trustworthy partners—they must examine and discuss this issue frankly.

Academic Integrity and the Values of Science

The issue of academic integrity is of such critical importance that I want to examine it in some detail. Universities must be willing to examine the foundation of scientific and scholarly research, the transmission of these values to succeeding generations, the contemporary pressures on researchers, and their responsibilities to research sponsors and themselves. Universities cannot make a convincing argument for more support and a stronger partnership without assuring that their own house is in order by doing everything possible to explain what it is that researchers and teachers do and by demonstrating that they are able to uphold the values they profess.

What are these values? What is the foundation of scientific and scholarly research that we hold so dear? The foundation is truth. As Jacob Bronowski noted in his book, *A Sense of the Future* (1977: 212–213):

> The end of science is to discover what is true about the world. The activity of science is directed to seek the truth, and it is judged by the criterion of being true to the facts. . . . We can practice science only if we value the truth. This is the cardinal point that has never been seen clearly enough, either by critics or by scientists themselves. Because they have been preoccupied with the findings, they have overlooked that the activity of science is something different from its findings. When we practice science, we look for new facts, we find an order among the facts by grouping them under concepts, and we judge the concepts by testing whether their implications turn out to be true to other new facts. This procedure is meaningless, and indeed cannot be carried out, if we do not care what is true and what is false.

Bronowski continues to discuss the values associated with this foundation. They include: independence and originality, a belief in the value of dissent, and an adherence to freedom of thought and speech. Central to these values is the importance of respecting another's point of view, even if it may be wrong. Ideas are interesting, even if they are disagreeable. Ideas and hypotheses exist to be debated, tested, proved, disproved, revised, built upon, or rejected.

These values make science—indeed, most scholarship—both an individual and a highly communal activity. They are also why science is said to be a self-correcting enterprise. Science works by a system of hypothesis, testing, and rechecking by alternate means in independent laboratories. In computer jargon, science is a fault-tolerant system. Like all human endeavors, it is not, and cannot be, completely free from errors. Over time, however, the system detects error and corrects itself. Thus the expectation of subsequent discovery certainly mitigates those forces that might tempt a scientist to cut corners or act with less than full honesty.

This is not to say that there are not abuses in the system. Bronowski himself alludes to such abuses when he talks about scientists becoming preoccupied with the findings over the process of research. The problem is real. And although I believe that academic misconduct and intellectual dishonesty are extremely rare—precisely because of the values held by scholars and the self-correcting nature of the system—there are nevertheless forces that push the contemporary researcher in directions that could prove undesirable.

The first is the increasingly rapid expansion of knowledge. The nearly instantaneous promulgation of research results by facsimile, computer network, mail, and frequent conferences is important for maintaining a worldwide community of colleagues informed and working within an up-to-the-minute framework. Ideally, such communication keeps the intellectual enterprise both current and coherent. However, it also contributes to an environment for working and thinking that is often frenetic. The shrinking of time made possible by modern communications can militate against the careful rumination and reformulation that might ultimately reduce error.

Second, the nature of incremental advances in some fields can push researchers into hasty work. Researchers and scholars often speak with disdain of "incremental" research and publication. However, in certain fast-paced fields where scientific mysteries are unfolding or new technologies are being developed or improved, progress is most often made incrementally, rather than by the revolutionary strikes of genius we all aspire to. Thus it is often the case that relatively modest advances—a new algorithm, a new chemical reaction, the effect of a new doping agent on a material's behavior, a new step in the sensitivity of a measurement technique—may have very real value but fleeting significance. This encourages rapid work, done by many different people, and a certain rush to produce and disperse results that also may reduce care, review, and reflection, thereby increasing the probability of erring or cutting corners.

Third is the culture of instantaneous news and fame. Although the popular press probably overstates this factor, it is quite true that scientists and scholars have egos. They are often highly competitive, a characteristic that generally works to the advantage of the advancement of science. However, the combination of scientific competition with the public's unquenchable thirst for sensation and its love of earth-shattering revelations creates temptations. The media are not to blame, because the draw of the spotlight and the appeal of recognition by one's colleagues is simply very human.

Finally, there is the opportunity for monetary gain. In recent decades, the time it takes to move from an idea or a discovery and into production of salable goods or services has shrunk enormously. Many of the boundaries between science and engineering have become fuzzy or disappeared altogether. Universities have become great engines of the modern economy and have increasingly worked together with profit-seeking partners. Indeed, federal, state and local policies have encouraged, and sometimes required, such activities. Through such interactions, university research has contributed significantly to the general welfare of the nation and world. But scientists would be naive to assume that their intimate ties with the corporate world are free from the possibility of conflict of interest, given the enhanced opportunities for personal financial gain.

Even with these pressures, however, I believe that intellectual dishonesty is extremely rare. Nonetheless, it does exist, and universities are obliged to ask what they can do to ensure that it becomes rarer still.

First, the scientific community must do a better job of strengthening and transmitting its system of values. The development of new knowledge and the education of students is at the very core of scientists' personal and institutional lives. However, scientists have a responsibility that is even more profound than the development and transmission of knowledge: to transmit to new generations the essence of scholarship and research, namely the methodologies and intellectual attitudes that demand the pursuit of truth with integrity and ethical rigor. Great teachers transmit and stimulate the passion, excitement, and beauty of intellectual endeavor. But it is equally important that they transmit and stimulate the meaning and the necessity of—and the passion for—intellectual integrity.

How can this be accomplished? The easy suggestion is to establish formal courses and perhaps require them of all students. But this is not necessarily practical or effective. Universities can, as I have asked MIT to do, establish broader mentoring of new colleagues—both faculty and students—and create occasional forums for addressing matters of ethics and integrity. Mentoring activities broadly and continually engage the university community in the matter at hand and, over time, develop in it an understanding of and value for intellectual integrity.

However, the greatest opportunities for education in this domain are individual and institutional actions. Values are reflected and publicized in actions. Regardless of what is said, the ways in which the members of the university community undertake their own scholarly activities, and the ways in which institutions deal with problems if they do arise, convey a lesson to all who are watching. The ways in which the community deals with academic misconduct or intellectual dishonesty also affect the level of society's confidence in and regard for the academic enterprise.

There have been great outcries for and against the policing of science. The response of the academic community must not be a knee-jerk defensiveness against the critics, but rather a plan for developing an effective method of self-governance regarding integrity in research. If universities are not able to do so, others will be only too glad to do it for them.

What universities do not need is more bureaucracy and increased overhead expenses for programs to enforce scientific integrity. What universities do need is thoughtful examination of the culture of academic research, as well as of the formal policies for handling cases of intellectual dishonesty or fraud. The National Academy of Sciences Panel on Scientific Responsibility and the Conduct of Research has suggested that a body external to academia could assist in writing model policies for investigating alleged misconduct and in keeping track of how much misconduct is reported and how it is handled (NAS 1992). Regardless of whether an external body is actually created, it is the responsibility of everyone in the university community to see that the taxpayer's fullest trust is earned and maintained.

The Academy panel has also pointed to some changes in the culture of universities that might prove beneficial, such as reducing the pressure on scientists to publish so many papers. Tenure committees, for example, might consider only a small number of a faculty member's best papers rather than placing so much weight on the number of publications.

But beyond the recommendations of the Academy panel, each university should review how it handles allegations of academic misconduct. At MIT, the provost and I asked a small group of distinguished faculty to do just that: to review and articulate our values in the conduct of academic research; to look at our own policies and procedures in light of those values; to compare these policies and procedures with federal and private guidelines governing research and to suggest revisions where appropriate; and to suggest creative ways of introducing mentoring and educational programs regarding both the conduct of research and the provision of broad career guidance throughout the entire academic community. I hope and expect other universities are engaging in similar activities. Universities need to turn the debate away from the front pages of the newspapers and to engage seriously with the thoughtful critics in addition to colleagues in designing policies and procedures to foster academic integrity and to deal forthrightly and equitably with problems when they arise.

Conclusion

These comments—both critical and laudatory—should be understood in the context of my belief that the university system in this country is the best in the world. This system is the result of a partnership that has been one of the great success stories in our nation's history and that is one of our most important investments for the future. Neither society nor the academic community can keep its flexibility, vigor, or quality by taking this partnership for granted. There are many reasons for the strains that do exist in this partnership. But there is also the interest and the will to renew the social contract between the American public and their research universities, and to forge an even stronger partnership among the universities, the public, and the government.

Source Note

Most of the material in this article is drawn from a lecture presented at Pennsylvania State University in the spring of 1991.

Bibliography

Bronowski, J. 1977. *A Sense of the Future: Essays in Natural Philosophy.* Cambridge: MIT Press.

Czujko, R., Kleppner, D., and Rice, S. 1990. *Their Most Productive Years.* Washington, DC: American Physical Society.

Kennedy, D. 1990. Influencing federal policy. Unpublished paper, Stanford University.

Lederman, L. M. 1991. *Science: The End of the Frontier?* Washington, DC: American Association for the Advancement of Science.

National Academy of Sciences (NAS). 1992. *Responsible Science: Ensuring the Integrity of the Research Process,* vol. 1. Panel on Scientific Responsibility and the Conduct of Research. Washington, DC: National Academy Press.

Office of Science and Technology Policy (OSTP). 1986. *Report of the White House Science Panel on the Health of U.S. Colleges and Universities.* Washington, DC: OSTP.

Rosenzweig, R. M. 1990. Building research partnerships. Keynote address at the 32nd annual meeting of the National Council of University Research Administrators, Washington, DC, November.

3

On Doing One's Damnedest: The Evolution of Trust in Scientific Findings

Gerald Holton

Science depends on trust. Confidence in other scientists' work is what makes the scientific enterprise cumulative. Without that trust, each scientist would need to replicate every relevant previous piece of research before building on it, qualifying it, or expanding it. Trust is thus integral to the existence of the community of scientists—what various observers have called the "union of eyes," the "republic of science," and even the "social contract" among scientists.

Trust is equally vital between the scientific community and the political community, for politics draws many resources from science, ranging from the ideological to the instrumental. The "republic of science" is often cast as a model of consensus and progressivity for the political community. Beyond that, the scientific community is productive and often bountiful in its provision of means to a better way of life. Yet because the public and their representatives are not scientists, they must trust scientists to conduct themselves and their community in an appropriate fashion.

In this chapter, Gerald Holton demonstrates that the trustworthiness of scientific findings is not a simple matter. His historical analysis creates a broader perspective within which allegations of academic misconduct can be better understood.

In essence, Holton argues against the naive notion that the job of scientists has always been simply to "report the facts." His account shows that the practice of science has been subject to evolution and change, so that the very idea of trying to report the details of an experiment is a modern one that would have seemed bizarre to some of the pioneers of experiment of the seventeenth and eighteenth centuries. Furthermore, he shows that the scientist is always confronted with decisions about which facts to report: the universe of information within which he or she operates is always far more vast than any single scientific report can encompass. And contrary to naive expectations, the facts can never "speak for themselves." What distinguishes great from pedestrian science is the creative scientist's capacity to recognize critical data amidst a sea of inconclusive observations. And that capacity has been subject to historical development.

The search for intersubjective (or "objective") understanding of nature's processes and laws, and the culture of science as a whole, are both under constant pressure for improvement. Even the handling of cases of alleged or proven scientific misconduct illustrate this point. For the credibility of scientific results does not have to depend on the probity of individual scientists. It is, rather, supported by the way the scientific community is organized. It is a community, unlike most others, where the shared ethos, developed over centuries, demands that every individual's claim is subject to open inspection, reexamination, and revision. It is also a community that has shown it can invent ways of dealing with new internal problems as they arise. The fragile bases for believing in, acting on, or financing the findings of science depend on the extent to which these inventions are regarded by the public and its representatives in the political world as sufficient and truly operative. [Eds.]

Introduction

Lest the title of this chapter mislead, I must declare at the outset that I do not address the scrutiny of scientists by politicians or the press, the topic currently so fashionable. I have done my share of writing on public trust in scientists, for the history of science does have something to say about it; one of this field's opportunities and obligations is indeed to trace changes in the intellectual and moral position of the community of scientists within society. Over the past decades, a rapid evolutionary process has been at work, shaping both the internal self-regulation and the external bureaucratization of the scientific profession, just as over the past four centuries the concepts, theories, and practices of science themselves have undergone a more familiar evolutionary development.

Indeed, a chief point I shall make is that the scientists' own methods for finding trustworthy research results are the result of a historic progression through stages, each with its own pitfalls and limits, and that here too we have no reason to think our ways will be adequate for tomorrow's needs. Our mental and physical tools have to develop constantly, if only because it usually becomes harder to take the next step as a field evolves, not only in the physical sciences, to which I shall refer here, but in every science. Hence there is within science a marked difference from one end of a century to the other in what it takes to trust one's own data or to trust the results of others. The same evolutionary process has external effects—among them how research is trusted by those outside the scientific community. Today's standards have been assembled from a steady stream of historically identifiable intellectual and social inventions; by the same token, we should expect the tools of science in the future to be in many ways quite different from today's. It will be wise to let the historical view prepare the mind for the next phase. To this end, I have selected a few telling episodes that illustrate the process of evolution of trust in scientific research results.

The Aesthetics of Necessity

An appropriate beginning is an example taken from the start of the modern period itself, which is usually associated with the work of Nicholas Copernicus. Until some years ago, the reigning opinion in the history of science was that the Copernican system was the very model of a new theory arising from a crisis brought about by the accumulation of data that contradicted an old and overly complex theory, in this case that of Ptolemy and his followers. Copernicus's book on the revolutions of the celestial objects, it was said, used more trustworthy observational data, yielded a better and more parsimonious theory, and so rescued the practitioners of the time by giving them at last a calculational method of greater accuracy for astronomical predictions.

But this interpretation is now seen by scholars in the field to be an ahistorical imposition of today's criteria of good behavior at the lab bench. On the contrary, Copernicus is chiefly an exemplar of the early introduction into science of essentially thematic presuppositions—that is, of deep convictions about nature on which the initial proposal and eventual acceptance of some of the most powerful scientific theories are still based (see, e.g., Gingerich 1975). Even if Copernicus's main goal had been to help the calculators of ephemerides, which was not the case, he gravely disappointed them. They were no better off after the publication of his *De Revolutionibus*, and in fact they continued to use the Ptolemaic system in essentially the form Ptolemy himself had set forth. For example, Copernicus's system of 1543 gave the same large errors for the predicted location of Mars—up to 5 degrees—as did the ephemerides of Regiomontanes in the 1470s.

Like Ptolemy, Copernicus selected just enough data (among them many with worse errors than he realized) to get his orbits, even bending some of them by a few minutes of arc as needed. But one must remember what he really wanted to achieve. This he made quite plain near the beginning of his great book: "to perceive nothing less than the design of the universe and the fixed symmetry of its parts" (1543: iii). What is more, he wanted to do so by sticking to what he called "the first principle of uniform motion" (that is to say, Aristotelian circular motion), instead of employing nonuniformity and nonconstancies as the Ptolemaics allowed. What convinced him to make this the cornerstone of his argument, and eventually persuaded his followers, was that he thereby produced a model of the planetary system in which the relative locations and order of orbits were no longer arbitrary but followed by necessity. In short, Copernicus is a case study of the privileging of an aesthetically based theory—above all the aesthetics of necessity—and of the temporary rejection of "data" that would appear to disprove a favored theory.

That daring approach has remained to this day a genetic trait in science, at least during the individual scientist's private, nascent considerations; and with all its dangers, the glorious thing discovered later was that it often works, and works without bending the data. Indeed, it can turn out that when a thematically and intellectually compelling theory is given a chance, better data, gathered with its aid, will eventually reinforce the theory. That is the meaning behind a remark Einstein made before the test of General Relativity: "Now, I am fully satisfied, and I do not doubt any more the correctness of the whole system, may the observation of the eclipse succeed or not. The sense of the thing is too evident." When a discrepancy of up to 10 percent remained between the first set of measured deviations of light and the corresponding calculations based on his theory, he responded: "For the expert this thing is not particularly important, because the main significance of the theory doesn't lie in the verification of little effects, but rather in the great simplification of the theoretical basis of physics as a whole" (Seelig 1954: 195). And even before more data came in that decreased the discrepancies, other scientists joined Einstein's camp, persuaded, in the words of H. A. Lorentz, that his grand scheme had "the highest degree of aesthetic merit; every lover of the beautiful must wish it to be true" (1920: 23–24). While this kind of procedure in science is of course much more risky for ordinary mortals, what most characterizes giants like Copernicus and Einstein as they struggle with problems too vast to be solved by the standard procedure, that is, induction from good data, is their intuition where science will go next. Scientific intuition, when it works, is a gift for which Hans Christian Oersted provided us the happy term, "anticipatory consonance with nature" (1966 [1852]: 450).

Yet even the giants of science can't count on their sense of "what's right." When Einstein asked himself what caused him to be so obstinately against a belief in the fundamentality of probabilism in physics, which the success of quantum mechanics forced on most others, he admitted to Max Born that he could not provide logical arguments for his conviction, but could only call on his "little finger as witness." By now we know with fair certainty that in this instance the "little finger" test failed.

Increasing the Probability of Reason's Claim

But I am getting ahead of my story. On the next stage along the line of the evolution of trust in one's results we find Galileo. Galileo's invention amounted to secularizing science, submerging the qualitative in favor of the quantitative as the earmark of truth, and elevating experimental checks from illustrations of the value of a theory to the test of its probability. In a famous passage in the Third Day of the *Two New Sciences* of 1638, Galileo's spokesman, Salviati, goes to such lengths to describe one of these experiments in detail—the accelerated motion of a bronze ball down an inclined plane, including his claim to have "repeated it a full hundred times"—that his supposedly skeptical listener Simplicio is made to confess why he trusts Salviati's account: "I would have liked to be present at these experiments; but feeling confidence in the care with which you perform them, and in the fidelity with which you relate them, I am satisfied and accept them as true and valid." To which Salviati quickly and eagerly responds, "Then we can proceed with our discussion" (1914 [1638]: 179).

This is still uncomfortably close to Aristotle's advice, in his *Rhetorica*, on how to persuade one's listener. Proof provided by the speaker is only part of it. Equally important, persuasion "depends on the personal character of the speaker, and on putting the audience in a right frame of mind." A few decades ago, the science historian Alexandre Koyré, unlike the more pliable Simplicio, found Galileo's account of the detailed care he took in conducting his experiments to be so overblown and unconvincing that he questioned whether Galileo had made any experimental checks at all. Instead, he argued, Galileo had presented only *Gedankenexperimente*, or thought experiments; he was not describing real ones at all (Koyré 1978 [1939]).

Fortunately for Galileo's reputation, some pages of his experimental tests have been found and analyzed, and they show that he—who once upon a time had been called the Father of Experimentation—did in fact do experiments. But modern analyses of his lab notes show why Galileo's private test results remained unpublished, and why he could not fully trust them to do more than, in his phrase, to "increase the probability" of "what reason tells me" (1914 [1638]: 170).

At that early stage of the evolution of trust, Galileo's private calculations were largely limited to proportions, and his published work was still narrative. Not until three decades later, in Newton's *Principia*, does the parsimonious style and axiomatic presentation, modeled on Euclid's geometry, take over—the sparse style of public science to which we are accustomed today. In Galileo's books there was still no use of algebra. Galileo does not announce his famous law of free fall as we do in elementary physics, $s = gt^2/2$. Rather, he makes the mathematically equivalent, but seemingly quite mysterious, statement: "So far as I know, no one has yet pointed out that the distances traversed, during equal intervals of time, by a body falling from rest, stand to one another in the same ratio as the odd numbers beginning with unity" (1914 [1638]: 153).

To have put it this way means that what counted for Galileo most was after all not the limited and perhaps rather silly case of a falling stone or a rolling ball, but the demonstration that terrestrial phenomena, of which these are examples, can be explained by the operation of integers—just as the Pythagoreans had dreamt (and as quantum physicists have proved in our century for atomic behavior). Galileo, too, was still engaged in a search for cosmic truths, a tendency which, for better or worse, had to be reined in as science evolved further.

Witnessing and Peer Review

The next steps in the evolution of trust were two. They came quickly, and both had to do with the ontological status of a "fact." While Galileo and Newton—a much better experimenter—could still regard their activities in the laboratory mostly as a private affair, others from the middle of the seventeenth century on struggled with the recognition that the door of the lab had to be unlocked; that Simplicio, so to speak, had to be invited to be present while the experiment was being performed. Fact had to be democratized. The academies and the Royal Society devoted much of their energies to the public elicitation of facts, the demonstrations of new phenomena carried on before an audience of interested fellow amateurs, acting as witnesses (Shapin and Schaffer 1989).

This practice can be still found a century and a half later, in the work of Oersted, who is most remembered for his publication on July 21, 1820 describing the discovery of the interaction of current-carrying conductors and magnets (Oersted 1820).[1] His hasty and perfunctory laboratory notes of findings, which changed all of physical science, would not be accepted now from a beginning student in an introductory physics course. Nor do they fit the new rules of explicitness that lawyers are busily writing today for government agencies to issue and for us to obey. And Oersted, a true Romantic, was still frank and personal in his publications; he did not know, as Louis Pasteur later told his students, that a scientist has to make the results "look inevitable."

Few modern researchers are likely to admit, as Oersted gladly did, that he had been completely convinced many years earlier of the existence of the effect he eventually discovered. (On the other hand, Oersted thus could in principle have been able to obey a memo of the sort sent out by our Secretary of the Department of Energy in 1989, asking that he be given "advance notice of all 'significant scientific discoveries.'") Oersted had been persuaded of a connection existing between electricity and magnetism by reading Immanuel Kant, who on metaphysical grounds proposed that all the different forces of nature are only different exemplifications of one fundamental force, a *Grundkraft.*

Oersted is also an example of a scientist still on the early rungs of the evolutionary ladder of trustworthy methods, as is demonstrated by his choice of procedure for what we now would call peer review. There being no other physicists of note available in Copenhagen at the time, the peers he selected to vouch for the truthfulness of his report were brought in more for presumed moral authority than for scientific acumen: he assured his readers that the experiments he reported were conducted in the presence of his friend Esmarck, the King's Minister of Justice, Wleugel, Knight of the Order of Danneborg and president of the Board of Pilots, and several other gentlemen whose word, one could assume, had to be trusted (1820: 273–274).

Deciding What Is a Fact

In the meantime, between Newton and Oersted, another enormously important problem was being wrestled with: how to determine which of all possible demonstrable events are indications of scientifically usable phenomena; which of them are really connected to the fixed regularities of nature, and which are merely passing phantoms, clouds with ever-changing form never twice the same, and thus reflecting only ephemeral concatenations? We might call it the problem of telling the difference between signal and noise.

The matter can be summarized most simply by contrasting the styles of research of Robert Boyle of the second half of the seventeenth century and of Charles DuFay in the early decades of the eighteenth. Boyle, while still proclaiming the superiority of reason over authority and even over experience, argues forcefully in his "Proemial Essay" for the inclusion in scientific reports of as many readings and as much detail of the experimentation as possible. He sounds quite modern when he favors "information of sense assisted and heightened by instruments" or argues that "artificial and designed experiments are usually more instructive than observations of nature's spontaneous acting." And he tries to respond to the warning of Francis Bacon: "Nothing duly investigated, nothing verified, nothing counted, weighed or measured is to be found in natural history; and what in observation is loose and vague is an information deceptive and treacherous."

But Boyle really doesn't yet know the difference between readings and data. His method of measurement is still quite primitive. For example, in his famous experiments on the compressibility of the air, he quietly assumes that the tube in which the gas column is being compressed is of uniform diameter, that the mercury in it is sufficiently degassed, and also that he can make, with high accuracy, naked-eye readings of its level with reference to a paper scale pasted on the outside of the glass tube.

Among Boyle's contemporaries there were a number of fellow amateurs of science and instrument makers who in fact dedicated themselves to increasing the range and accuracy of basic measurements. They typified the sort of person who, from the sixteenth century to our day, has reveled in the invention of more and more ingenious devices for measuring time, distance, angle, or mass. Their achievements have made it possible to obtain more accurate and useful values both of derived quantities (such as force, pressure, electric charge) as well as of the physical constants of nature. That quest is itself a rather heroic chapter in the history of science and technology. It recounts such triumphs as the nearly logarithmic rates of decrease in the uncertainties of weighing (three orders of magnitude, from about 1550 to 1950); in the error, in seconds per day, of time measurement (eight orders of magnitudes in three centuries, from about 1650); and in the error of astronomical angular measure (five orders of magnitudes, from 1600 to the 1920s). The same story is told in the nearly logarithmic rate of increase of the resolving power of microscopes (from 0.9 to 0.2

microns, between 1840 and 1880) and of the energy reached for initiating elementary particle reactions (ten orders of magnitudes, from about 200 KeV in 1930 to over 1,000 TeV in the mid-1980s).[2]

Taking that route toward greater reliability of results was not in the forefront of Boyle's thought. But more significant for us here is that he was still wide open to anything that might happen before his eyes and was apt to count everything observable as a fact to be used in his research. In his desire to find the gold of trustworthy detail, he was smothered by slag and mere sand. And that is of course one reason why not he but one of his readers, Richard Townley, discovered among Boyle's data what we now call Boyle's law (Conant 1967: 62).

With Charles DuFay, as Lorraine Daston has pointed out, we see a representative of this new stage of development. In his researches on electricity, DuFay alters his experimental conditions constantly, with the aim of isolating the relevant variable. He is interested only in those facts that are characteristic of a large class of bodies, not of isolated species, and can therefore be organized into some scheme or simple rule. In short, he intuits the modern characteristic of "facticity": that fact is consensual, invariant, and universal. And he signals the necessary next step toward the modern base of trust in experimental reports: he takes on a collaborator, not merely to witness but to try to repeat what DuFay has observed. DuFay realizes that experiments are very difficult to do right at an early stage of a science. More generally, he insists that phenomena do not assume the status of fact until other investigators go over the same ground, repeat the observations, and give their consent. This practice was a key invention along the path of increasing trust in research results.

The Emergence of Teams

If we had enough space, we would now look at the next contributions to the development of trust as seen in the work of Oersted's followers, particularly Michael Faraday and James Clerk Maxwell. But let me go on to one of Maxwell's successors at the Cavendish Laboratory, J. J. Thomson, usually identified as the discoverer of the electron in the 1890s, and to his student and successor, Ernest

Rutherford. They exemplify something new in the evolving tale of trust within the lab.

With "J.J.," then still in his mid-thirties, as the new professor, the Cavendish Laboratory officially became a graduate school in 1895. Ernest Rutherford was among the first advanced students to arrive at the lab, soon to be followed by others of great talent, including a few guest researchers from abroad, even from the United States. What occurred there was another step toward the modern phase of science: the forging of a group identity among the participant scientists.

Even before J.J. established himself as a major charismatic leader, he had his students assist him in his experimental researches and left it to them to chase down the details once he himself was satisfied he had obtained the right order of magnitude from theory. But in addition to his natural way of mentoring his assistants and being interested in each student's work, he had another, rather English weapon: regular afternoon teas for all researchers in his room. That was an occasion to develop the sense of community among the researchers, who not only shared the same interests but also depended on one another in their work. By and by, there emerged what Dong-Won Kim (1991) in his recent doctorate thesis identified as a core research group around the professor, though only rarely was the collaboration with J.J. so close as to merit, in his view, joint authorship of resulting papers.

The whole operation, though successful, was still a hand-to-mouth affair, financed largely by student fees and scholarships. Only inexpensive materials, such as glassware blown by the students themselves, were likely to be used. The sociology of the modern laboratory was beginning to come into view, but not its economics. However, the combination of the selection of the advanced students and the socialization of the group as a whole around the more and more famous central figure was so effective that it gave rise to a kind of moral imperative, a bonding. These individuals would mind no hardship and, I should imagine, would rather have committed ritual suicide than betray the scientific norms of the group and of the times. They were not separate, career-oriented passers-through but "citizens" of a community who shared in a common spirit.

Rutherford built on this system with a vengeance when he started his own operations at McGill University, then at the University of Manchester, and finally at the Cavendish Laboratory in Cambridge. Lab work was now planned more carefully, progress was discussed with each student at least weekly, and a list of research tasks that Rutherford thought could be completed with reasonable success was given to each student annually. Almost as a byproduct of his campaign to understand the atomic nucleus, Rutherford's approach was an early example of a style that would soon lead to real team research. Each student still had to make the most of his or her own equipment, and each had been selected by Rutherford with an eye to being "first class," willing to work hard as if in a constant race and with only the most minimal funds. But despite the fact that they all labored on different projects that were linked only in Rutherford's mind, he expected full loyalty to himself and to the laboratory. He was ever-present. Some of his students later reported they felt he had adopted them. Not for nothing was he known to his students as "Papa," according to J. C. Crowther. Trust in each other and in the quality of their data was assured as automatically as it would have been in a healthy family or an isolated tribe of hungry hunters.

Still, the organization of the laboratory was by no means the modern one of team research, that is, research where a whole laboratory, with many participants, works together on a shared project. Nor was the Cavendish operation, while magnificent in its production of major results, "modern" in terms of understanding the relative value of money for apparatus versus the extra time spent on work when equipment is scarce. I once asked James Chadwick, who had discovered the neutron after a painstakingly long search in 1932 at the Cavendish with the most primitive equipment, to describe the atmosphere there at the time. He wrote back that his hope had been to settle the question whether the neutron existed by building a small accelerator. However, "no suitable transformer was available and, although Rutherford was mildly interested, there was no money to spend on such a wild scheme. I might mention that the research grant was about £2000 a year, little even in those days for the amount of work which had to be supported. I persisted with the idea for a year or two. . . . I had quite inadequate facilities, and no experience in such matters. I wasted my time—but no money" (1971 [1962]: 28).

Wasting a year or two of a Chadwick for lack of budgeted funds was soon to become unacceptable, and so would a style of research that failed to take full advantage of the combined talents of a group. The earliest exemplar of that new style in physics was the "family" of young researchers that Enrico Fermi educated and assembled around himself with spectacular results from about 1929 to the mid-1930s. Here, again, group loyalty was extraordinary. This being Rome, and Fermi being thought infallible in quantum physics, his students called him "Il Papa." Money was no real problem; with excellent political instinct, Fermi had secured the patronage of the Italian state for his laboratory needs. And with equally good managerial instinct he designed methods for deciding on promising research lines and then pursuing them fiercely, with a division of labor within the group that could be a model even now.

All this is implied even in Fermi's decision to list as authors of his laboratory's publication every one of the various members of that team—at least in physics a hitherto unheard-of public assignment of credit and sharing of responsibility for the results. Consider the paper reporting the startling discovery of the artificial radioactivity produced with beams of slow neutrons, sent to the *Ricerca Scientifica* on October 22, 1934. This paper can be said to have effectively opened the nuclear age by announcing the discovery of resonance and introducing the concept of the moderator for nuclear reactions, and it figured in the award to Fermi of the Nobel prize for physics four years later (Fermi et al. 1962 [1934]).

There were five authors, given in this order: Fermi, Amaldi, Pontecorvo, Rasetti, and Segrè. We know from their various biographical statements that the young associates of Fermi worked closely under his inspiration, and the listing of authors, with the name of Fermi put first, is an indicator of their relationship. But I must add that when I looked in the archives in Pisa at the lab notebooks the group kept during that project, I found it difficult to disentangle from those laconic entries who did what, and when. Fermi's innovations, in that more innocent age, did not go so far as to foresee the kind of self-protective bookkeeping forced upon labs in our time by the expectation that (as in the case of a researcher in the laboratory of David Baltimore) even the Secret Service might be called in by eager congressional investigators who doubt the trustworthiness of reported scientific results.

As we know, the invention of publicly assigning and distributing responsibility in works that have multiple authorship has been successful beyond all reasonable expectations. Records are being broken year by year as larger and larger groups launch themselves against harder and harder problems. A paper in the *Physical Review D* of June 1, 1992 lists some 365 authors from thirty-three institutions on three continents (Abe et al. 1992). Such indicators signal a kind of phase change in the social matrix of science, to which we shall have to turn in closing. Before that, however, we must at least briefly note another crucial milestone on the road to greater reliability of findings: the adoption of statistical methods for the analysis of data.

The Statistical Treatment of Data

An example from the first decade of this century will underline how recent, from a historical perspective, are the details of data processing. These issues, such as the use of statistics and even of significant figures, are now so common that many assume they have been with us since the beginning of time. For this example I return to Robert A. Millikan, a scientist whose use of data in his unpublished lab notes I have had occasion to write about (Holton 1978); I found it a revealing case study of the way a creative researcher often exercises judgments during the nascent, private phase of experimental work that may look impermissibly arbitrary when examined with the benefit of hindsight. This time I want to look briefly at Millikan's publication of February 1910 in the prestigious *Philosophical Magazine*, his earliest "big" paper on the road to what two years later turned into his triumphant "oil drop" method for measuring the charge on the electron, *e* (Millikan 1910).

In 1910, at age forty-two, Millikan was still essentially unknown, and he was near despair about his chances to break through to scientific prominence. He had no way of knowing that the February 1910 paper, on a "new modification" of a well-established method of using the motion of water droplets in electric fields to find *e*, was to point him soon in the right direction. But the paper allows us to glimpse how, at that point in early twentieth-century history, one could still treat one's data.

In the section entitled "Results," Millikan frankly begins by confessing to having eliminated all observations on seven of the water drops—these, for various reasons, he decided had to be "discarded." A typical comment of his, on three of the drops, was: "Although all of these observations gave values of *e* within 2 percent of the final mean, the uncertainties of the observations were such that I would have discarded them had they not agreed with the results of the other observations, and consequently I felt obliged to discard them as it was" (1910: 220). Today one would not treat data thus, and one would surely not speak about such a curious procedure so openly.

Millikan then presented his results in tables, in each of which he had gathered the data from one of six series of experiments, and gave also the raw calculations they yield. Almost every detail there is astounding from our present point of view. Each observation carries Millikan's opinion of its likely reliability: "The observations marked with a triple star are those marked 'best' in my notebook. . . . The double-starred observations were marked in my notebook 'very good.' Those marked with single stars were marked 'good' and the others 'fair'" (1910: 221–223). Correspondingly, he assigns to each of the six series a weight to be used in averaging all results to obtain a final value; no details are given, but by inspection one sees that in Millikan's mind he correlates nine or ten stars in a series with a weight of 7, seven stars with a weight of 6, five stars with a weight of 4, three with 3, and zero with a weight of 1.

Within each of the six series in the tables, there are more surprises. There is little attempt to keep the significant figures straight during calculations; no reason is given why the readings are clumped the way they are, except that each clump corresponds to a set of drops that are guessed to carry the same number of electron charges. To find a value for *e* within a clump of observations, Millikan uses the average of all the individual data for the voltages needed to balance the various drops and of the times of fall when the electric field is off—instead of calculating the charges drop by drop and then applying statistical data reduction for a final result. Throughout, an air of utter self-confidence pervades the paper. But one must add one more fact: Despite all these "peculiarities," the final charge of the electron obtained with the new method ($e = 4.65 \times 10^{-10}$ e.s.u.) was excellent for its day, and could not be improved for many years. Oersted's phrase, "the anticipatory consonance with nature," comes again to mind.

Statistics had entered theory in the eighteenth century and experimental science by the mid-nineteenth century. But it took much longer to get general agreement on the proper use of statistical analysis—for example, to deal with such bothersome problems as outriders in lab data, and when or how to "reject" observations; for in the real world of the laboratory, unlike its idealization by many nonscientists, one must be prepared to find even that "at some level, things will happen that we cannot understand, and for which we cannot make corrections, and these 'things' will cause data to appear where statistically no data should exist. . . . The moral is, be aware and do not trust statistics in the tails of the distributions" (Bevington and Robinson 1992). Only in the mid-twentieth century did good practice on those points become general at the level of the ordinary lab worker. When excellent books such as E. Bright Wilson's *Introduction to Scientific Research* (1952) became available, every student could read what "reasonable procedures" for data acquisition and reduction might be, including under extreme or difficult conditions.[3]

"Big Science"

But of course the landscape has changed immensely in the decades since then. Our journey brings us now to the latest and most difficult stage, that of the gigantic thought- and work-collectives in science. The story is told well in Peter Galison's book, *How Experiments End* (1987). Chapter 4 is devoted to the discovery in the 1970s of so-called neutral-current events, such as the observation of neutrino-electron scattering. The detection of these events led to the confirmation of the theory of the unification of the electromagnetic and weak nuclear forces by Sheldon Glashow, Steven Weinberg, and Abdus Salam—a modern triumph in the *Grundkraft* program of Kant and Oersted.

The theoretical questions and fierce debates concerning the possibility of such a unification took an entirely new turn when, at the big accelerator at CERN in Switzerland, a photograph of a single electron event was found by a group from Aachen in January 1973. The detecting instrument they used had been given the name Gargamelle, after the mother of Gargantua. It was a monstrous bubble-chamber device, holding 12 cubic meters of liquid propane—a major engineering project of its own, and a symbol of the interpenetration of engineering and science in our century. Thousands upon thousands of pictures are exposed in the course of an experiment and then are given to the team trained to scan the negatives. In this run, one exposure was recognized as being strange and new by one of the scanners. She passed it into the hands of a research student, who identified the traces as those of electrons. The following day the student carried it up along the complex hierarchy of a big group to the next rung on the ladder, to the deputy group leader. Thinking it of considerable interest, he in turn brought it to the institute director, who later called it a picture-book example of what they had been expecting for months: a candidate for neutrino-electron scattering.

The crucial point now was to assess the background, that is, the probability of a masking event. The director took the picture to England, to a fellow expert. And so on, and so on. The circle of belief expanded constantly. But still, this was only one picture, a single event, and not even an event one could immediately reproduce. What a huge difference from the days of DuFay and all his followers, all the way to Fermi's group! Yet, as Galison writes: "Experienced bubble chamber experimentalists found the Aachen electron particularly compelling. Their specialty was famous for several critical discoveries grounded on a few well-defined instances. The omega-minus [particle] had been accepted on the basis of one picture, as had the cascade zero. Emulsion and cloud chamber groups had also compiled arguments based on such 'golden events,' including the first strange particles, and a host of kaon decays" (1987: 183).

Still, too much was at stake to publish immediately. One might believe the evidence, but one could not yet believe in it, trust it to be reproducible in principle. One now had to calculate the probabilities of all kinds of other reactions that could masquerade as the one assumed to be happening. Background events had to be ruled out; if one of those was what the exposure was about, it wouldn't test the Glashow-Weinberg-Salam theory at all.

There ensued an agonizing set of discussions within the large collective, as drafts of publications were being debated. We see here the next step in the evolution we are tracing: the internalization within a large group of fellow workers of the array of old but public methods of arriving at trustworthy results—debates among fellow specialists, preparation of publications and refereeing, and all the other safeguards of standards of demonstration. It is as if a new, large organism were doing science within its own boundaries. Finally, Galison concludes, "the members of the collaboration persuaded themselves that they were looking at a real effect. So it was that no single argument drove the experiment to completion. . . . It was a community that ultimately assembled the full argument" (1987: 193–194).

The brief publication of the final result in the *Physics Letters* announcing the finding had boiled down years of work into a few sentences and neat graphs—a far cry from the extensive documentation of Kepler and Boyle (Hasert et al. 1973). A heterogeneous group of subcollectives from Belgium, Britain, France, Germany, Italy, Switzerland, and the United States had come to the decision to stake their reputations on a new kind of physics. And it is worth noting that the new finding had been distributed to the network of fellow specialists long before the mailing of the July 19, 1973, issue of the journal, through prepublication "preprints."

Over the past few decades preprints have become in some branches of science by far the preferred method of "publishing" and of keeping up with the "literature"; the Physics Department library at Harvard alone receives more than 4,000 preprints per year from more than 500 universities and other institutions worldwide. These in turn are now made quickly available through an electronically accessible database specially designed for them. Here, too, we catch a glimpse of the spectacular rate of change in the way science is being done. The next frontier is already in view: A debate is in progress among physicists at large colliders whether "outsiders" should have access to as yet unpublished data, even before the preprint stage.[4]

But to return to the CERN result. The story did not quite end there. By November 1973, a group at Fermilab in Illinois had gathered enough evidence that all this work at CERN was a mistake—that no neutral current existed. They were almost ready to announce their finding. But then they found traces of the effect that "would not go away," and in April 1974 they too published their evidence in favor of the neutral current.[5]

One lesson of this story is that just as one can trace the influence of Bach and Beethoven in some of Schoenberg's music of the early twentieth century, so can one find some habits, methods, and standards of the old days in the mega-teams of modern science. But the new conglomeration is also a transformed entity, and it deals with a transformed science. In many branches, the more difficult it is to find credible access to the phenomenon, the more dependent the scientist is on apparatus built by others, on data gathered by still others, and on calculations carried through by yet others. When one is immersed in a large group, some members of which might change from month to month, it is not easy to know whom to believe, when to believe it, and how well to believe it. And the ground is shifting once again before our eyes. More than 900 physicists had signed on to a single experiment on the planned Superconducting Supercollider in Texas before its cancellation,[6] and the number of collaborators on the Human Genome Project is greater still.

All these changes raise a vexing question. Where lies the source of trust—for a participant in the research as well as for the reader of the publication—when most individuals within a team of widely diversified competences cannot vouch for every aspect of the published results, when perhaps not even one person in the whole group can be expected to be fully conversant with every element? A bitter rule of thumb has evolved: whereas credit for success tends to be parceled out unequally (those close to the field usually feel they "know" who in a long list of authors had the best ideas or most novel techniques), an embarrassing error or misconduct by even a single participant is likely to bring discredit to all the members of the team. Thus there is now a big incentive to invent ways to distribute partial responsibility early and proportional justice, when necessary, later. For example, some research papers identify the contributions of individual contributors to a complex group effort. But because of the differences in location, or the wide spectrum of subspecialties involved, or the styles of leadership, etc., there is such a variety in the way teams are run that no single set of rules is likely to emerge. We are watching an experiment in self-governance, analogous perhaps in some ways to the transition that took place when the Pilgrims arrived in the New World and set about the task of forming a new society.

Doing One's Damnedest, Revisited

Still, some fundamental things apply as time goes by. It was my luck to have known, and to have learned my trade under, one of the last of the physicist-philosophers, P. W. Bridgman—a Nobel Prize winner (1946) for establishing the field of experimental high-pressure physics almost single-handedly, and also the father of the movement in philosophy of science known as operationalism. Operationalism centers on the position that the meaning of a concept is in its measurement or other test. In the 1930s Bridgman made a famous, useful, and very operational statement, usually remembered as: "The scientific method is doing your damnedest, no holds barred" (1955: 535).[7]

For him personally, doing one's damnedest meant total absorption, from 8 a.m., when he arrived at the lab by bicycle, rain or shine. It meant incredible productivity: Bridgman published some 230 substantial scientific papers that came to seven volumes of his collected scientific works (Bridgman 1964), as well as several books on the philosophy of science. It meant unceasing dedication to the process of getting things clear in his own mind. His style was completely "hands on," every datum taken by him or by his one, dedicated long-term assistant. Most apparatus built by his own hands. Only two or three joint papers (one with a young chemist named J. B. Conant). Very few thesis students—only those whom he could not persuade otherwise. Only a few hundred dollars a year needed for materials. No overhead. No paperwork. Paradise!

New Demands on Scientists

Bridgman's statement is still a great definition of how to get at trustworthy scientific results. But in the intervening decades the milieu has so evolved that we need to supplement that advice, to bring it into the 1990s. At the very least we must add a few words:

> While doing your damnedest, watch how your presuppositions are holding up; make sure you understand the results of other people along the chain on whose work you are relying; and keep in view that more and more of the findings of science, resonating through society, have additional results far beyond those sought initially.

Bridgman himself discovered that last point in his own way. He once thought that science is essentially "value free"; but with the rise of Nazism in the 1930s, he saw that German scientists were no longer able to act as free intellectuals, and were in many cases coopted as servants of a Fascist state. Therefore he published a "Manifesto" in 1939, to the effect that he was closing his lab to visiting scientists from totalitarian countries (Bridgman 1939). The act served as an early, widely noted reminder for scientists in the United States that their work has ethical dimensions even beyond what Jacob Bronowski identified as the principle that binds society together, in science as well as outside, the "principle of truthfulness."

Most scientists today face not only new conditions, such as their complex relationships with sophisticated instruments and their collaborators. They also face a different kind of fact, one that increasingly asserts itself: By obtaining the ever more necessary support from government or industry, they have incurred new obligations, not least the responsibility that their research results be sharable far beyond their own labs. For reasons having little to do with the relatively rare case of scientific error, misjudgment, or misbehavior, new pressures for accountability are jostling them from all directions. All parties involved are still fumbling around a bit—the National Institutes of Health, Congress, the universities, the scientific societies. They are struggling to invent new methods for preventing and defending against charges of misconduct, real or imagined—just as, at an earlier stage of evolution, they had to learn about factuality and significant figures.

This new task, too, is one which history has thrown their way. Scientists cannot let others create the new realities for them. The next phase in the continuing evolution and transformation of the methods of science will depend on the actions of today's researchers. At stake are the problem choices for tomorrow's scientists, their rights to respect and support, the attraction of their fields for future scientists, and even the ancient hope that scientific thought itself is an exemplar for attaining trustworthy conclusions.

Notes

1. This is the English translation of Oersted's pamphlet, *Experimenta circa Effectum conflictus electrici in acum magneticam.*

2. For some of these estimates, see Turner (1967); on the astronomical angular measures, see Pledge (1959: 291).

3. See Porter (1986) for a guide to the development of the concern with statistics in science.

4. See, for instance, Hernandez (1992). On federal courts ordering the "sharing" of confidential data, see Marshall (1993).

5. After accepting the existence of neutral currents, one of the chief investigators wrote: "Three pieces of evidence now in hand point to the distinct possibility that a [muon]less signal of order 10% is showing up in the data. At present I don't see how to make these effects go away" (quoted in Galison 1987: 235).

6. See Flam (1992).

7. The passage from this source reads: "The scientific method, as far as it is a method, is nothing more than doing one's damnedest with one's mind, no holds barred."

Bibliography

Abe, F., et al. 1992. Limit on the top-quark mass from proton-antiproton collisions at $\sqrt{s} = 1.8$ TeV. *Physical Review D* 45:3921–3948.

Bevington, P. R., and Robinson, D. K. 1992. *Data Reduction and Error Analysis for the Physical Sciences.* 2nd ed. New York: McGraw-Hill.

Bridgman, P. W. 1964. *Collected Experimental Papers.* Eds. H. Brooks, F. Birch, G. Holton, and W. Paul. Cambridge: Harvard University Press.

Bridgman, P. W. 1955. *Reflections of a Physicist.* New York: Philosophical Library.

Bridgman, P. W. 1939. "Manifesto" by a scientist. *Science* 89:179.

Chadwick, J. 1971 [1962]. Some personal notes on the search for the neutron. In *The Project Physics Course Reader*, 25–31. New York: Holt, Reinhart & Winston.

Conant, J. B., ed. 1967. *Harvard Case Histories in Experimental Science, Case l: Robert Boyle's Experiments in Pneumatics.* Cambridge: Harvard University Press.

Copernicus, N. 1543. *De Revolutionibus Orbium Coelestium.* Nuremberg, f. iii (v).

Fermi, E., Amaldi, E., Pontecorvo, B., Rasetti, F., and Segrè, E. 1962 [1934]. Influence of hydrogenous substances on the radioactivity produced with neutrons—I." In E. Fermi, ed., *Collected Papers*, vol. 1, 761–762. Chicago: University of Chicago Press.

Flam, F. 1992. Big physics provokes a backlash. *Science* 257:1468.

Galilei, G. 1914. *Dialogues Concerning Two New Sciences.* Trans. H. Crew and A. de Salvio. New York: Macmillan.

Galison, P. 1987. *How Experiments End.* Chicago: University of Chicago Press.

Gingerich, O. 1975. "Crisis" versus aesthetic in the Copernican Revolution. In A. Beer and K. A. Strand, eds., *Copernicus, Yesterday and Today*, 85–93. Oxford: Pergamon Press.

Hasert, F. J., et al. 1973. Observation of neutrino-like interactions without muon or electron in the Gargamalle Neutrino Experiment. *Physics Letters B* 46:138–140.

Hernandez, O. 1992. Letter. *Science* 258:13.

Holton, G. 1978. Subelectrons, presuppositions, and the Millikan-Ehrenhaft dispute. In G. Holton, *The Scientific Imagination*, 25–83. Cambridge: Cambridge University Press.

Kim, D. W. 1991. *The Emergence of the Cavendish School: An Early History of the Cavendish Laboratory, 187l–1900.* Unpublished Ph.D. dissertation, Harvard University.

Koyré, A. 1978. *Galileo Studies.* Trans. J. Mepham. Atlantic Highlands, NJ: Humanities Press.

Lorentz, H. A. 1920. *The Einstein Theory of Relativity.* New York: Brentano's.

Marshall, E. 1993. Court orders "sharing" of data. *Science* 261:284–285.

Millikan, R. A. 1910. A new modification of the cloud method of determining the elementary electric charge and the most probable value of that charge. *Philosophical Magazine* 19:209–228.

Oersted, H. C. 1820. Experiments on the effect of a current of electricity on the magnetic needle. *Annals of Philosophy* 16:273–276.

Oersted, H. C. 1966 [1852]. On the spirit and study of universal natural philosophy. In L. Horner and J. B. Horner, trans., *The Soul of Nature.* London: n.p.

Pledge, H. T. 1959. *Science Since 1500.* New York: Harper.

Porter, T. M. 1986. *The Rise of Statistical Thinking, 1820–1900.* Princeton: Princeton University Press.

Seelig, C. 1954. *Albert Einstein.* Zurich: Europa Verlag.

Shapin, S. and Schaffer, S. 1989. *Leviathan and the Air Pump: Hobbes, Boyle and the Experimental Life.* Princeton: Princeton University Press.

Turner, G. L'E. 1967. The microscope as a technical frontier in science. In S. Bradbury and G. L'E. Turner, eds., *Historical Aspects of Microscopy*, 175–197. Cambridge: W. Heffer & Sons.

Wilson, E. B., Jr. 1952. *An Introduction to Scientific Research.* New York: McGraw-Hill.

4

Integrity and Accountability in Research

Patricia Woolf

In 1981, after a small number of instances of falsified research had publicly surfaced, Representative Albert Gore, Jr. (D-TN) and Senator Orrin Hatch (R-UT) each held hearings that inquired into scientific misconduct and the ability of the research enterprise to control misconduct on its own. At the Gore hearing, Philip Handler, then president of the National Academy of Sciences, called the problem "grossly exaggerated" and assured Gore's oversight subcommittee that the existing system "operates in an effective, democratic and self-correcting mode." The issue persisted at a low level of concern in Congress, the funding agencies, and research institutions for several years until, in 1987, accusations surfaced that a paper co-authored by Nobel laureate David Baltimore of MIT contained falsified research. Other oversight subcommittees, most notably those chaired by John Dingell (D-MI) and Ted Weiss (D-NY, now deceased), began their own investigations of this and other episodes of alleged misconduct and highlighted examples of ineffective and undemocratic operation and failures of self-correction by the scientific community. Quite simply, the relationship between the political community and the scientific community has not been the same since.

In this chapter, Patricia Woolf considers how research misconduct has exacerbated even the usual difficulties of science policy by undermining public trust in scientists and scientific institutions. Although Woolf acknowledges that there are other forces at work both in science and in the broader society that make scientific fraud particularly "fashionable," she rejects the contention that the problem of research misconduct is overstated and should not be engaged publicly. The threat of failing to engage in a public dialogue is that research misconduct could be the proverbial last straw: when the pursuers of scientific truth—who have been trusted as neutral arbiters to resolve disputes among interested parties—turn out themselves to be deceitful, the camel's back of public trust is broken.

Woolf attributes much of the attention to research misconduct to the "conventions of deception" that surround us in politics, religion, sports, and the media as well as in science. Because it also emphasizes persuasion rather than presentation, science is in this way of a piece with the broader culture of advocacy and advertising. Woolf believes that the scientific community can rise above these other spheres of society to improve scientific research and education. Although she begins with a list of the burdens that the vagaries of scientific integrity add to the fragile contract, she concludes with a set of practical guidelines to restructure the relationship between society and science. [Eds.]

Introduction

In the best of times, science policy is daunting. It was baffling enough before there were issues about charlatans misusing public funds for falsified or plagiarized research. Consider these nine difficulties:

1. Research is an unpredictable enterprise that frequently succeeds best when it achieves ends that are not foreseen.

2. It depends on a highly trained and individualistic workforce that values personal autonomy as a prime source of creativity, and is ideologically convinced of the ultimate social value of its endeavors. In the academy, there is a fundamental belief that research succeeds most when individual scientists choose to do the research that each feels is important. However, in practice, autonomy is often secondary to a hierarchical system in which disagreement is seen as insubordination.

3. The community sustains a belief that science cannot succeed without research scientists enjoying "the right to be wrong." Moreover, scientists claim for themselves the singular qualification of deciding what is right and what is wrong.

4. The system that integrates the results of research—publication in scholarly journals, professional meetings, scuttlebutt, e-mail,

corridor talk, and telephones—is loosely managed by scientists themselves. It is very little studied, not well understood, and totally unregulated. Scientific publishing is presided over by a mixture of commercial and academic interests that are not accountable in the short run to any of the interested parties—not to the funding agencies, not to research institutions—and only in the long run to the scientist readers and authors.

5. The productivity of the research enterprise is usually measured not by meeting predefined goals, but by a surrogate system: papers published in scholarly journals.

6. The goals of science are purportedly driven by abstract ideals, but are often justified on the bases of medical, social, military, and commercial benefits.

7. Results are often expressed in language that is incomprehensible to those outside of the particular scientific discipline, including the legislators who authorize its funds.

8. There is no official certification or licensing system for scientific research, with the possible exception of the doctoral degree. There are no rating agencies, no balance sheets, and few annual reports. Performance audits are rare, but do exist in some multicenter clinical trials.

9. The research workforce has numerically outgrown its traditional institutions. The new constellation of research institutions consists of publicly and privately funded, for-profit and not-for-profit sectors, that may or may not be affiliated with traditional institutions such as universities, hospitals, and corporations. That is, the research enterprise is radically pluralistic.

In the best of times, getting the public to support research would be a pretty hard sell. And these are not the best of times.

Trust

Science is supported on trust; indeed, the entire enterprise operates on trust. This vital trust has been jeopardized by fraud and misconduct in research.

The coordination of disparate activities within science to produce reliable knowledge depends on a tactical balance among trust, skepticism, and negotiation. Sociologists from traditionalists such as Robert Merton to revisionists such as Bruno Latour have addressed how that system works for scientists themselves. Historian of science Gerald Holton (this volume) describes how that system has evolved over time.

The great majority of U.S. citizens has accepted that scientific research is good and honorable in addition to being beneficial. But this acceptance is granted on faith since, as Dorothy Nelkin (this volume) reminds us, only a tiny percentage of the public are both knowledgeable and concerned about scientific matters.

Disenchantment

Over the past fifteen years or so, however, there have been a number of developments in which scientific progress has proved to be much more complicated than anticipated. In the eyes of some, the benefits of science have been offset by perplexing dilemmas and unwholesome developments. Many of these complications have become highly politicized and are now part of national debates in which scientific evidence is summoned to provide corroboration (or merely decoration) for strongly held and divergent views that are apparently immune to reason. Science is being challenged not only for its failures, but also for its successes, because of the vast social changes they have engendered.

Attention focuses on science for several reasons. Traditional issues concerning human obligations have been revolutionized by new technologies and the legal apparatus that has evolved as a result. For example, assumptions about the beginning of life have been challenged. Thanks to sophisticated contraception, sex with little chance of reproduction has become commonplace. The legal system is trying to cope with novel means of reproduction: in vitro fertilization has opened the challenges of postmenopausal births, surrogate mothers, and frozen embryos. Fetal tissue research—supplied with tissue from abortions—has been joined to these issues by critics.

The end of life has also been irreversibly altered, posing enormous problems for the science, medicine, and technology that have prolonged and redefined it. Death (as we once understood it) is no longer final. Dying has become less an act of God than a predictable and sometimes manipulable stage of life.

Animal rights activists attacking research (in addition to furriers and meatpackers) have created controversies that, at their extreme, often involve nothing less than reestablishing the place of humans in the order of things.

AIDS has brought many of these issues into sharp relief. In the face of a frightening epidemic, there are telling criticisms of the focus, pace, funding, and priorities of research and of the regulation of therapies resulting from that research.

Hazardous substances and unperceived emanations are associated with scientific and technological progress. As specialists argue over alar, dioxin, radiation, the greenhouse effect, PCBs, EMR, rDNA, and CFCs, the public is often left with baffled anxiety and an increased dependence on expert scientific opinions. When expert opinions change, confusion can lead to disillusion, even when opinions evolve for reasons that scientists feel are good and sufficient (Marshall 1991).

Trouble

Science is perceived to be in trouble, epitomized by the *Time* magazine cover story "Science Under Siege" (Jaroff 1991). Compounding the trouble over the substance of science is concern about the practice of science, that is, about cases of cheating in research. Arguments about the definition of research misconduct continue to affect policy. The current definitions almost always include fabrication (making up the results of research without doing experiments), falsification (rigging experiments, substituting photographs, graphs, gels, isotopes, etc.), and plagiarism (putting one's own name on someone else's research, taking what modern society calls their "intellectual property" without attribution).

Some definitions of research misconduct also include attributing one's own work to someone else by putting the name of someone more prestigious on the research. There is also a lawyerly catchall in which misconduct is also defined as other practices that seriously deviate from commonly accepted standards.

Speaking at Harvard University, Benjamin Lewin, editor of the journal *Cell* (which published the disputed paper in the "Baltimore affair"), expressed his surprise that fraud in science had received so much attention: "I find it ironic that this should occur at a point when science is, in fact, more successful than it has ever been before, when research is advancing more rapidly than before." His explanation was "that the scope of the problem [of research misconduct] has been vastly overstated" and that "much of what has been said represents a basic misunderstanding of how science works" (Lewin 1989).

There are other possible explanations for this level of attention to fraud in science, such as: the increased attention to science in the press and the newsworthiness of scandal; the zeal of members of Congress, and more particularly of their staff members eager to find fame by finding waste, fraud, or threats to public safety; the rotten apple hypothesis; a changing research environment characterized by a pressure to publish and a shortage of funds—even for highly ranked research; the megalomania of scientific superstars and the envy of their competitors.

An alternative explanation does not preclude a partial role for these factors, and speaks directly to Lewin's sense of irony. A recent work on forgery and plagiarism in the Middle Ages argues that "forgeries and plagiarisms . . . follow . . . fashion and can without paradox be considered among the most authentic products of their time" (Constable 1985 [1983]: 2). Charlatans are attracted to the dominant activities of their period. Scandals in research are thus—paradoxically—affirmative evidence that science is successful, a socially central enterprise that can be considered on a par with politics and government, finance, and popular religion—areas where there has been no shortage of charlatans. Furthermore, since biomedical research is currently one of the most dynamic and productive areas of research, it is no accident that most misconduct has occurred in research related to medicine.

In fact, public attention necessarily focuses on failed deceivers: the ones who have been caught. "The secret of successful forgers . . . is to attune the deceit so closely to the desires and standards of their age that it is not detected or even suspected at the time of creation" (Constable 1985 [1983]: 1). So, perhaps a useful consequence of fraud in science can be the opportunity to examine what these failed attempts at scientific deceit tell us about the desires and standards of our age and the ways they are reflected in scientific research.

We live in an era of pervasive misrepresentation and a certain vagueness about the importance of truth; and scientific research perforce is conducted in that atmosphere. New words and phrases blur the distinction between fact and fiction. Orwellian neologisms such as "docudrama," "advertorials," "adversorials," and "infotainment" publicize this fuzziness. There are "news releases" that are thinly disguised advertisements. From the Reagan White House we learned of "plausible deniability." Political campaigns produce "infomercials." Agencies generate "disinformation." On Cable News Network, "factoids" are fill-ins that report items that may be true, but are so trivial that they do not matter much. Nowadays, pictures do lie as digitizing techniques can remove and insert persons or buildings, and change the seasons of the year.

Parents and churches still teach their children to tell the truth. But part of socialization, even maturity, in contemporary society requires the ability to understand the conventions of deception. For instance, academics are expected to present their ideas in their own words, but politicians routinely deliver speeches that others have written. Estimates for remodeling a kitchen are less than you will eventually be expected to pay. Estimates for insurance claims are expected to be dishonestly high and adjusters sometimes discount claims, anticipating overstatements by claimants. Automobile insurance costs include a certain (high) percentage in anticipation of faked claims.[1] The system amounts to a share-a-charlatan kind of "equity." In universities, traditional scholars who believe calmly in factual evidence, and for years have trusted that conclusions based on evidence were—somehow—true, now find that they have colleagues who question everything and have turned both evidence and interpretation into a spicy gumbo to be sampled only at risk of intellectual heartburn.

From business to sports to international politics, there are attitudes that competitiveness requires one to bend the rules; and that pushing the limits of regulation is just another example of creative thinking. An example from basketball illustrates: "I don't say break the rules. I do say every set of rules has a certain flexibility factor, and if you're always thinking about that factor and how it can work in your favor, it will" (Auerbach and Dooley 1991: 46).

But the public and its representatives expect science to be different. First of all, the fundamental commitment of science is to truth. As modern technology, and social and legal thought about humanity and its place in the universe have made much traditional thought anachronistic, people turn increasingly to scientists not only for facts, but for guidance. Yet this search for guidance occurs at a time when doubts have been raised about whether scientists can be trusted as responsible agents.

At the same time, public perceptions linking science to technology compound suspicions. Negligent practices have resulted in the waste of scarce resources. Even the appearance of carelessness can undermine confidence in science. Certainly the most dramatic example was the Challenger explosion. But the public does not know how to assess necessary and unnecessary risks. After all the promises, what does one think of a Hubble telescope that needed contact lenses? Contradictory statements by experts undermine faith in science, as have reappraisals of the dangers of Love Canal and asbestos in building materials.

The misuse of public funds for unauthorized or personal purposes undermines confidence in science, as does the publicity about government challenges to university overhead charges paid for by research contracts and grants. Some of these violations may be trivial fiscally, but they are material in their effects on public trust. Ski vacations on federal grants, scientific conferences held in ritzy spas, and boondoggles disguised as continuing medical education (often sponsored by pharmaceutical companies) provoke the cynicism even of the participants. They lead to a sense of extraordinary and possibly unmerited privilege.

While scientists tout the importance of peer review as a quality control measure, on occasion they have subverted peer-reviewed processes to gain advantage for their own projects. Line item budget requests and political intervention in research funding decisions may gain temporary advantages for individual institutions, but damage faith in professional judgments. There are journals that purport to be peer reviewed, but many are reviewed only in part, or by in-house editors. Other journals are frankly available for pay, and even legitimate peer-reviewed journals publish sponsored symposia that promote products and are often less carefully reviewed (Bero, Galbraith, and Rennie 1992). One university subpoenaed confidential peer reviews from a scholarly journal (in history) for use in sexual discrimination litigation; surprisingly, the journal supplied the reviews.[2]

The increasing use of scientists as partisan representatives of conflicting viewpoints in litigation has also contributed to disenchantment and cynicism about scientific expertise. How can antagonistic experts both be right? If they cannot, how can opposing scientists each be objective, rational and trustworthy? Who will decide which experts are qualified?[3] In modern science the disinterestedness of scientists has been linked to their objectivity and thus to the reliability of their research. Although most people recognize that scientists are unlikely to be completely neutral with respect to the results of their studies, they are skeptical about scientists who appear as advocates for certain positions rather than as objective presenters of facts (Proctor 1991). In several allegations of research misconduct, there have been charges that apparent financial conflicts of interest have distorted the knowledge base on which other decisions depend. Since subsequent research and public policy decisions and, ultimately, human health may be jeopardized, considerable attention has focused on allegedly inaccurate claims made in the clinical testing of drugs (Wingerson 1983; Begley 1988).[4]

In addition to concern about financial conflicts of interest, there is worry about conflicts of commitment in universities. Some teaching failures are attributed to the diversions of research, or the suspicion that faculty spend too much time jetting to conferences held in glamorous resorts. The quip—paraphrasing Thorstein Veblen—is that this represents "the leisure of the theory classes." Parents have objected to incomprehensible accents of foreign-born graduate student teachers who replace faculty in some courses in research universities. A story, perhaps apocryphal, tells of a state legislator in the Midwest introducing a bill requiring that all faculty members in state universities take an examination in spoken English because his daughter had difficulty understanding her math teacher. The math teacher turned out to hail from Brooklyn. University tuition increases have outpaced inflation at a time when many Americans are suffering from the long recession. No one seems to know where the money goes: professors are not all that rich. But the new laboratory buildings springing up on campuses add to speculation that research is the culprit.

Although researchers depend on the community of scientists for validation of results, it seems that the public is increasingly focusing on individuals as prize winners, entrepreneurs, and spokespersons for various interests. Genuine heroes attract public attention and approbation for science; but when they fail, those failures are newsworthy, and the calamity is far greater than it would be if they had not been idolized in the first place (Hilts 1992).[5]

The theatricality and haste associated with the initial announcements of "successful" cold fusion did not enhance the stature of science as a careful enterprise—certainly not in the short run. The ballyhoo and the prompt responses of other scientists demonstrated the strengths and the weaknesses of peer review and replication as warrants for reliability in science, and in the long run may increase respect for community criticism. Cynics will note, however, that in the initial announcements researchers, university administrators, and state officials were unconstrained by scientific caution as they battled for intellectual property rights before establishing whether the phenomena they were arguing about really existed (Close 1991).[6] Careful physicists who worked on fusion did so in an atmosphere tainted by the overselling of a dubious discovery.

Popular articles, including several in scientific journals, about the discovery of the AIDS virus and about the Gallo-Montagnier controversy also cite the press conference at which Secretary of Health and Human Services Margaret Heckler announced that Gallo's lab had discovered the probable cause of AIDS and announced that a vaccine was only a few years off.

Speaking to the point that forgery follows fashion, the successes of science have attracted highly publicized imitators. Not everything that is publicly touted as research is really scientific. A dolphin research center in Florida operates on a not-for-profit basis, but admits it does not really have any scientists on the staff.

In an article entitled, "The Wizards of Hokum," the author referred to Biosphere 2 as "not the only project to blur the line between hokum and hard science; in fact a vital symbiosis seems to be developing" (Toufaxis 1991).

Research Fraud

In this litany of public concerns, the most threatening is deliberate misrepresentation, including fabrication, falsification, and plagiarism. In spite of all the publicity and numerous reports from scientific organizations, the problem of research fraud and misconduct remains particularly refractory because we still have no empirically based idea of how significant it is in terms of the impact on new scientific knowledge. Many scientists continue to think it is largely a political problem or, as one official at the National Institutes of Health (NIH) put it, based on misunderstanding. Data available from public sources reinforce that view. In 1991, the Office of Scientific Integrity of the Public Health Service reported that after investigations of 174 allegations had been completed, in fewer than twenty instances had there been a finding of misconduct and sanctions imposed (NAS 1992).

Public agencies resist giving any but the most sanitized summaries of completed cases of misconduct. In the absence of systematic and reliable data, scientists have difficulty assembling information that could provide a basis for improved public policy. Details have often been provided only by news reporters or by those whose actions and credibility have been challenged in connection with particular cases. Of the few persons who are interested in determining how significant the problem is (in terms of frequency and impact), some appear to have a separate antiestablishment agenda. Dorothy Nelkin (this volume) describes the moralistic tone of critics outside of science. The same is true within science.

The only people who can intervene effectively or creditably in dealing with allegation of misconduct are scientists themselves. They will do so only if they think the problem is real and serious. The public good is not well served by enforcing standards that produce careful but pedestrian science. Yet there is a need to offset the images of negligent research practices and defensive posturing. What will persuade scientists that malfeasance in research is serious now and potentially disastrous for the pursuit of knowledge, and for the institutions that are devoted to learning and teaching?

Policy and Politics

Science policy should address the interests of citizens vis-à-vis science, asking what the public interests in the oversight of science are, and whether they are well reasoned. For the sake of scholarly discussion, the analysis should be dissociated from feelings about those who have taken an interest in science fraud and their often undignified pursuit of publicity.

Prior to 1992, with the impasse that had been created by a Republican executive and a Democratic Congress, members of Congress found their opportunities to legislate very limited. That left them and their thousands of staffers considerably greater incentives to investigate and regulate than to legislate. Despite the confrontations and accusations, defensiveness and posturing, congressional hearings on research misconduct have addressed issues that are important for science policy. There is food for thought in the transcripts, and video tapes of the hearings could be an important pedagogic tool for the responsible conduct of research. To what end, and in what terms, has Congress questioned scientists?

Health and Safety of the Public

Representative Ted Weiss (D-NY, now deceased), chairman of the Subcommittee on Human Resources and Intergovernmental Relations of the House Committee on Government Operations, focused on falsification, plagiarism, hasty and negligent research, and financial conflicts of interest. He asked the deceptively simple question, "Are scientific misconduct and conflicts of interest hazardous to our health?" (U.S. Congress 1988a).

Waste, Fraud, and Abuse

Representative John Dingell (D-MI) is the chairman of the Subcommittee on Oversight and Investigations of the House Energy and Commerce Committee. His bold and energetic staff provide a stream of subjects for oversight. It is perhaps cold comfort for scientists, but it is important to realize that scientific research is not the subcommittee's only target. Oversight responsibilities have led to intensive investigations of the financial auditing profession, the insurance industry, and defense procurement. Sometimes it appears that the committee's investigative template is simply being switched from one subject and fitted to another (U.S. Congress 1988b, 1989, 1990a). Indeed, the responses of the financial auditors' Treadway Commission look very much like documents that have been drafted by various scientific groups simultaneously testifying to the adequacy of current controls and urging reforms that will strengthen them.

Science as a National Resource

The Subcommittee on Investigations and Oversight of the House Science, Space and Technology (formerly chaired by Robert Roe, D-NJ, now retired) focused on science as a crucially important resource, an American natural treasure (U.S. Congress 1990b). Reliable science is needed for the economy to stimulate innovation in industry and restore American competitiveness in the global marketplace. The integrity of the biomedical research programs at NIH is vital for understanding diseases and developing cures. To science and technology are entrusted the task of contributing to almost the entire agenda of government action, from restoring the national infrastructure to developing environmentally sensible materials and sources of energy; from increasing the quality and quantity of food to developing weapons and other tools of war and security.

Responses to Misconduct

As the congressional hearings revealed, not much good can be said about scientific fraud. But bungled investigations at some research institutions gave the impression that the disease could not be as bad as the cure. Unfortunately, the public (including the scientific public) has heard more about inept inquiries than about efficient and effective ones. In addition to charges and countercharges, an amazing amount of apologetic quibbling has been reported. Is plagiarism less culpable in review articles than in original research? In grant proposals rather than publications? When it is spoken rather than written? Scientists whose work depends on the privilege of making errors (and correcting them) at public expense must distinguish errors from fraud and not equivocate—not even at the urging of lawyers.

Scientists' Responses

Most institutions now have policies and procedures to deal with allegations of misconduct (AAMC 1992). In response to an NIH requirement, institutions that apply for training grants have instituted some formal training in the responsible conduct of research. The National Academy of Sciences Panel on Scientific Responsibility and the Conduct of Research has issued a major, two-volume report (NAS 1992; NAS 1993).

Most scientists now recognize that Congress must oversee the way funds are spent, although they object to the managed publicity of some of the hearings and deplore competition among committees and their staffs. Many do not understand the functions of the research oversight offices in the Department of Health and Human Services or the Office of the Inspector General at the National Science Foundation. Those who sympathize with the problems of these agencies still have little good to say about the attempted or proposed solutions.

University Responses

Research misconduct has been a crucible for testing the autonomy and the responsibility of research faculty, and the relationships of scientists to university administrations, funding agencies, and the public at large. Because scandals are predictably newsworthy, the confrontations have often taken place in public.

Whereas earlier appraisals of the misconduct situation focused on how much fraud in science existed, academic leaders now realize that a relatively small number of incidents have in fact disproportionately damaged public confidence in science. The problem is serious, and not all that rare. Misconduct can happen at any institution and experience shows that the better the institution, the more likely it is to happen, or at least be detected (Woolf 1988). There are no indications that research misconduct is going to be less serious or less frequent in the near future.

University administrations and leaders in the profession, partly under the stimulus of granting agencies, have been guiding (and pressuring) the faculty to be individually concerned. And yet faculty who have not been directly confronted with a case of misconduct are largely unsympathetic to the issue. The large majority of scientists who are honest and conscientious are resentful or (at best) ambivalent about being forced to fix something that they feel is not broken. Faculty members who have been embroiled in controversies are frequently defensive; because of the constraints imposed by confidentiality most are unwilling or unable to share their experiences with others. Faculty have, nevertheless, been reluctantly involved in developing procedures to handle allegations, and now face the task of improving research teaching and mentoring.

In the past, university administrations reacted defensively, awkwardly, and slowly. They are now better prepared, but still greatly challenged by conflicting demands for prompt resolution of allegations and deliberate procedures that are fair to the accuser, the accused, and—most important—to the public that supports research. They are caught in the squeeze between demands that universities "do something" and the understandable desire of the faculty not to be pushed into more rules, committees, and paperwork.

There are obvious limitations to university procedures and explicit rules for research. Procedures and rules will certainly not prevent all fraud, and they may even lead to an increased number of allegations. As rules become more explicit, violations will become more evident. It is also possible that determined deceivers will find more subtle ways to challenge the spirit if not the letter of the law. But in addition to being required by law, institutional guidelines provide a rational framework for dealing with situations that can be intellectually complicated and emotionally charged.

Research misconduct has put scientists on notice that research training can and should be improved. In today's heterogeneous research environment, scientists cannot assume that all participants in laboratory life share the same values or the same standards. In particular, the concept of intellectual property is not well understood. Now that the possible financial benefits of research are considerable and more promptly realized, scientists and their administrators need to spend more time on understanding the distribution of rewards. Similarly, standards are evolving in the area of conflict of interest. Scientists need to lead in understanding and persuading others how the profits from scientific and technological expertise should be distributed.

There is evidence that deviant behavior can evolve, and that individuals who succeed in minor transgressions can progress to more significant ones. Institutions can limit the rewards for opportunistic practices and thereby shape attitudes as well as behaviors. In the short run, the goal should be to foster sound research practices, rather than to make moral persons. In the longer run, responsible actions in themselves can influence values.

Universities and funding agencies and their oversight offices will continue to face serious pitfalls and painful choices. Several lessons of research misconduct are obvious in retrospect:

The personality or status of the person who makes an allegation must not be confused with the seriousness or merit of the charge.

The professional reputation of the accused and assertions of innocence will not substitute for prompt, thorough, independent inquiry.

Scientists have long memories. The problem does not go away if the accused person goes away. As tempting as that solution seemed to be in the past, it is not acceptable to let researchers resign when suspicions are raised.

Lawyers and scientists approach problems differently. Both approaches are needed in determining the merits of allegations. The role of university counsel must be defined; the interests of the university and of faculty members represented by university lawyers can diverge as investigations progress.

The disclosure of the results of investigations—to whom, and when, and how much information—remains a major concern, especially when the prompt correction of the scientific literature collides with due process considerations for the accused.

The problems are worse and the solutions must be more delicate when the relevant research information is sensitive or classified (for instance, when confidential records of research subjects are involved).

Failures to observe requirements of due process have exacerbated the tension between faculty scientists and administrations, and between research agencies and universities. Inquiries, investigations, hearings and legal proceedings have dragged on unconscionably long—in some cases, five or six years. The problem of how scientists deal with their careers during those periods is a problem not only for the individuals concerned but also for universities and for the public. Does innocent until proven guilty apply to the continued use of scarce research resources?

In sum, scientists have enjoyed remarkable freedom in the United States to conduct investigator-initiated research, to select and teach their own successors, to award public monies to themselves, and above all to set their own standards—including the rare privilege of making and correcting their own errors. The price of that freedom is individual and professional vigilance.

Notes

1. See J. Romano, "A state crackdown on insurance fraud," *New York Times* (December 27, 1992), section 13: 1, 6.

2. See S. K. Dervan, "Rice responds to motion for retrial," *The Rice Thresher* (September 2, 1991):1.

3. The Supreme Court granted certiorari in *Daubert v. Merrell Dow Pharmaceuticals, Inc.*, and heard the case, which in part involves the criteria that expert testimony should meet, in fall of 1992. In June 1993 the Court issued its opinion that judges, rather than juries on the one hand or scientific institutions such as journals on the other hand, should determine the appropriateness of expertise. See *Daubert v. Merrell Dow Pharmaceuticals, Inc.*, 113 S.Ct. 2794 (1993).

4. See also J. Foreman, "A battle over research ethics that touches every parent," *Boston Globe* (February 17, 1992):Science Watch, 1.

5. See also J. Foreman, "Glory and greed on the trail of the AIDS virus," *Boston Globe* (March 23, 1992):Health Science, 25.

6. See also W. Broad, "Cold fusion still escapes usual checks of science," *New York Times* (October 30, 1990):C1.

Bibliography

Association of American Medical Colleges (AAMC). 1992. *Beyond the "Framework": Institutional Considerations in Managing Allegations of Misconduct in Research.* Washington, DC: AAMC.

Auerbach, R., and Dooley, K. 1991. *MBA: Management by Auerbach.* New York: MacMillan (as excerpted in America West's *Airline Magazine* 7(2), April 1992).

Begley, S. 1988. Fraud in the laboratory? *Newsweek* (April 11):69.

Bero, L. A., Galbraith, A., and Rennie, D. 1992. The publication of sponsored symposiums in medical journals. *New England Journal of Medicine* 327:1135–1140.

Close, F. 1991. *Too Hot to Handle.* Princeton: Princeton University Press.

Constable, G. 1985 [1983]. Forgery and plagiarism in the Middle Ages. *Archiv für Diplomatik* 29:1–41.

Hilts, P. J. 1992. The science mob. *The New Republic* (18 May):24–31.

Jaroff, L. 1991. Crisis in the labs. *Time* (August 26):45–51.

Latour, B. 1987. *Science in Action.* Cambridge: Harvard University Press.

Latour, B., and Woolgar, S. 1979. *Laboratory Life.* Beverly Hills: Sage.

Lewin, B. 1989, June. Remarks at the Symposium on Ethics and Research, 100th Anniversary of the Graduate School of Arts and Sciences, Harvard University, Cambridge.

Marshall, E. 1991. A is for apple, alar, and . . . alarmist? *Science* 254:20.

National Academy of Sciences (NAS). 1992. *Responsible Science: Ensuring the Integrity of the Research Process,* vol. 1. Panel on Scientific Responsibility and the Conduct of Research. Washington, DC: National Academy Press.

National Academy of Sciences (NAS). 1993. *Responsible Science: Ensuring the Integrity of the Research Process,* vol. 2. Panel on Scientific Responsibility and the Conduct of Research. Washington, DC: National Academy Press.

Proctor, R. 1991. *Value Free Science?* Cambridge: Harvard University Press.

Toufaxis, A. 1991. *Time* (September 31):66.

U.S. Congress. 1988a. *Federal Response to Misconduct in Science: Are Conflicts of Interest Hazardous to Our Health?* House of Representatives, Committee on Government Operations, Subcommittee on Human Resources and Intergovernmental Relations. 100th Cong., 2nd sess., September 29. Washington, DC: U.S. Government Printing Office.

U.S. Congress. 1988b. *Fraud in NIH Grant Programs.* House of Representatives, Committee on Energy and Commerce, Subcommittee on Oversight and Investigations. 100th Cong., 2nd sess., April 12. Serial No. 100–189. Washington, DC: U.S. Government Printing Office.

U.S. Congress. 1988c. *Scientific Fraud and Misconduct and the Federal Response.* House of Representatives, Committee on Government Operations, Subcommittee on Human Resources and Intergovernmental Relations. 100th Cong., 2nd sess., April 11. Washington, DC: U.S. Government Printing Office.

U.S. Congress. 1989. *Scientific Fraud.* House of Representatives, Committee on Energy and Commerce, Subcommittee on Oversight and Investigations. 101st Cong., 1st sess., May 4, 9. Serial No. 101–164. Washington, DC: U.S. Government Printing Office.

U.S. Congress. 1990a. *Maintaining the Integrity of Scientific Research.* House of Representatives, Committee on Energy and Commerce, Subcommittee on Oversight and Investigations. 101st Cong., 1st sess., April 30 and May 14. Washington, DC: U.S. Government Printing Office.

U.S. Congress. 1990b. *Maintaining the Integrity of Scientific Research.* House of Representatives, Committee on Science, Space, and Technology, Subcommittee on Investigations and Oversight. 101st Cong., 1st sess., June 28. No. 73. Washington, DC: U.S. Government Printing Office.

Wingerson, L. 1983. Biotechnologist faked results in race for patent. *New Scientist* 97:1339.

Woolf, P. 1988. Deception in scientific research. *Jurimetrics Journal* 29(Fall):67–95.

5

The Public Face of Science: What Can We Learn from Disputes?

Dorothy Nelkin

Science and technology policy in the United States has always been a domain of elites, largely scientists themselves but also civilian and military leaders who sought to harness scientific productivity for national goals. Clearly associated with the technological triumphs of World War II and therefore with winning the Cold War as well, science was an area that the American public readily entrusted to its delegates for administration. For a generation after World War II, science was a relatively uncontroversial means to relatively uncontroversial ends.

Beginning in the 1960s, however, science became controversial, both as a means of achieving goals like peace, prosperity, and social justice, and in its association with the excesses of such ends as industrial capitalism and containing Communism. Without agreement on the directions or uses of science and technology, the public was no longer willing to delegate decision making in science policy as freely as before. The peace movement, the environmental movement, the women's movement, and other mass social forces created a demand for public participation in scientific decision making.

But as Dorothy Nelkin argues in this chapter, the character of scientific controversies has changed since these early movements. The issues at stake have been transformed from being overtly political to being overtly moral. In Nelkin's view, a fundamental ambivalence about science and technology lies at the center of the nation's scientific controversies. One component of this ambivalence is a respect for scientists, a desire for instrumental technology, and a belief in the neutrality of science. The opposing component is a skepticism of elites, a tendency toward moralistic politics, and a reliance on the clash of interests to resolve disputes. Over time, the kind of disputes involving science have moved from involving questions of representation (who will decide—the experts or the people?) to questions of morals (are these activities appropriate to conduct?). Nelkin is not optimistic about the resolution of such rights-based, moral conflicts, although she holds out some hope that negotiation, mediation, and other participatory strategies can close the gap at least temporarily. Beyond that, she argues, the fragile contract between society and science is dependent on the resolution not of specifically scientific disputes, but of broader political tensions in U.S. society. [Eds.]

Introduction

There is growing interest these days in the public face of science—stimulated perhaps by the wrinkles or pockmarks caused by political pressures from Congress, conflicts over misconduct in research, the threat of regulation over the internal affairs of science, reduced funding, and an increasingly critical press. Professional societies are examining their relationship to the media. Research communities are increasingly interested in science education, fearing that the young who see the face of science are turning away. The Human Genome Project has allocated over 3 percent of its budget to studies of the social and ethical implications of genomic research. Survey researchers are investigating public attitudes toward science.

Understanding the public face of science is not a straightforward matter. Attitude surveys tell us that the public has generally retained a certain faith in science as important to social progress, but they tell us little substantively about the diversity of public views, and they quickly reveal that most people know very little about science until it affects their immediate interests. The media provide another window on the public face of science, but the press tends to convey the findings of science as breakthroughs, disembodied from their methodological and policy context (Nelkin 1994). Moreover, the images of science in the press, often shaped by scientists themselves, are interpreted by readers or viewers according to their biases.

Public controversies over science offer a view that reflects the concerns of specific interest groups. People are not much involved in the details of science policy and priorities except as these bear on specific issues or causes that touch on their personal concerns. But once an issue becomes controversial, it is quickly brought to the attention of lawmakers and regulators. In this way, controversies influence the relationship between science and government. I draw on material from studies of disputes to suggest patterns in the evolution of controversies over the past few decades and to explore what they reveal about the public image of science.[1]

Such disputes concern many different aspects of science and involve many different publics. For example, an animal rights movement has obstructed the use of long-accepted research methods. Antiabortionists blocked the availability of federal funds for fetal research, until President Clinton lifted the ban in 1993. Some feminists have questioned the development and use of new reproductive technologies. Gay rights activists are challenging the scientific procedures and guidelines that have delayed the availability of new therapies for AIDS (Acquired Immune Deficiency Syndrome). Religious groups with moral reservations join farmers with economic concerns to question certain biotechnology applications. Corporate interests have objected to dietary guidelines established by scientific commissions. And of course whistleblowers generate troubling disputes by encouraging congressional investigation of the research process.

Science controversies concern more than just the protection of immediate interests: some are struggles over meaning and morality; others over the distribution of resources; and still others over issues of power and control. Science has become, in effect, an expression of social tensions—an arena to battle out deeply contested values in U.S. society.

The prevalence of disputes over science is in some ways a measure of its success. As science becomes an increasingly important aspect of public culture, it is inevitably subject to scrutiny, criticism, and debate. But recent debates must also be understood in the context of the long history of public ambivalence toward science. I first lay out some sources of this ambivalence; then, drawing on specific cases, I examine some issues that have generated recent controversies and suggest how they have evolved. Finally, I explore the meaning of disputes in terms of contested public values and their bearing on the political role of scientists and the routes to resolution.

Public Ambivalence

Attitude surveys indicate little change in the general level of public support of science over the past twenty years (Miller 1990). When questioned, most people say they perceive both science and technology as instrumental in achieving important goals. Yet disputes over science and technology have proliferated since the early 1970s. At first they focused on concerns about the environmental and health effects of technology, mobilizing community activists directly affected by technological decisions and engaging scientists as experts providing technical support (Fischer 1990). By the late 1970s, however, a number of protests focused directly on scientific research, as animal rightists, antiabortionists, and whistleblowers challenged the longstanding autonomy of science from political regulation and public control.

Many scholars have addressed the significance of these trends. One example was the 1978 conference at the Massachusetts Institute of Technology (MIT) on "The Limits of Scientific Inquiry," which examined the proposition that some kinds of research should not be done at all (Holton and Morison 1979). Throughout the 1980s there was talk about the "crisis" in science, as research seemed faced with attacks by forces from both the left and the right. By 1990, in *The Descent of Icarus,* Yaron Ezrahi suggested that these mounting attacks against science represented a major conceptual change in the role of science in society: "In the closing decades of the twentieth century, the intellectual and technical advances of science coincide with its visible decline as a force in the rhetoric of liberal democratic politics" (1990: 13).

Attacks on science are far from new or recent. Current disputes reflect a long history of ambivalent public attitudes toward science in U.S. society (Mazur 1981). The acceptance of the authority of scientific judgment has long coexisted with mistrust and fear, revealed, for example, in early opposition to innovations such as vaccination or to research methods such as vivisection. But controversies today also reflect the increased scale of science and its pervasive influence on contemporary life. Science these days represents both progress and peril; while welcomed for its benefits, it is feared as a source of risk (Krimsky 1982). People are aware not only of the economic cost of research, but also of its social cost to research subjects—humans, fetuses, animals. They express ambivalence over whether the benefits of science outweigh the costs, and whether the promises of science can compensate for the risks. And while the public still views scientists as a source of neutral knowledge, and therefore as exempt from political accountability, there is also considerable mistrust in the ability of scientists to regulate themselves (NSB 1989).

In following disputes over the past decade and a half, I have observed some significant changes. The controversies that began to emerge in the early 1970s tended to focus on questions of political control: who controls decisions about the development and applications of science? The questions represented, in part, the so-called "crisis of authority" that prevailed in political life of that time, as the antiestablishment values originating in the critiques of the 1960s spread to all institutions, including science (Salomon 1977). These critiques also inspired political activism, animating local interest groups to mobilize against decisions that affected their lives.

By the end of the 1980s, however, protesters were framing their attacks on science in moralistic and absolutist terms. Antiabortionists claimed that fetal research was "wrong" and should be abandoned regardless of its clinical benefits. Animal rights activists deemed animal research "immoral," to be banned regardless of its contribution to medical knowledge. Critics of science—creationists, antiabortionists, feminists, ecologists, animal rightists—were not just weighing risks against benefits; they were expressing their concern about instrumental activities that transform the status of nature, fetuses, women, or animals from ends to means, resources or tools. They began to pose fundamental challenges to the instrumental reasoning of science; and they have radicalized many of the protests that began in the 1970s.

From the viewpoint of many scientists, the activities of protest groups resemble nineteenth-century Luddism—the wholesale rejection of scientific and technological change that Brzezinski once called "the death rattle of the historically obsolete" (1970). From the perspective of activists, however, protest is a positive force, necessary to counter the misuse of science by major social institutions that activists believe will reduce humans as well as nature to commodities. Those who challenge science often see themselves as preserving the moral values lost in the course of technological change.

The proliferation of controversies today suggests a growing polarization in the public image of science—a polarization between those who see science as a communal activity essential to social progress and those who see it as an instrumental activity driven by economic interests (Richards 1988); between those with programmatic agendas seeking to implement specific goals and those with moral lenses concerned about accountability, responsibility and rights. Some controversies remain mainly at a policy level where science policy issues are debated by experts, ethicists, and policy elites. But increasingly, others are public protests—even social movements—driven by moral outrage and assuming an increasingly strident tone.

The Issues in Dispute

The most intense and intractable disputes over science concern research practices that seem to have direct moral implications.[2] The practice of animal experimentation has revitalized the nineteenth century antivivisection movement under the rubric of animal rights, which has become a large, well-funded, and very angry social movement built on the moral premise that animals should not be exploited as a resource for human use, that is, they should not be "test tubes with legs." The more extreme activists deny all moral boundaries between humans and other animals, insisting that all species should be treated the same way (Jasper and Nelkin 1992). Advances in fetal tissue transplantation have met similar moral objections from antiabortionists who regard the fetus as an innocent and exploited victim (Maynard-Moody 1992). And the creation of transgenic animals through techniques of biotechnology has inspired protests from those convinced that such manipulations are a violation of the natural order (Krimsky 1991).

Disputes like these reflect the pervasive concern with morality in U.S. society. Even as biomedical research brings about dramatic improvements in medical care, critics question—indeed, try to stop—certain areas of science that threaten their convictions. While some defend disputed practices for their therapeutic benefits, others see only abuse. For critics, the use of women as surrogate mothers or the use of animals or fetuses in research is morally unacceptable, threatening concepts of personhood and violating nature. Understandably, they call for the cessation of such research.

A second type of controversy focuses on the information offered by scientists involved in the ubiquitous disputes over health hazards (Rosner and Markowitz 1991). Gaps in scientific information leave considerable leeway for conflicting interpretations. Inevitably, uncertainties about the extent and the nature of risk result in conflicts among scientists, which in turn aggravate public fears. Thus scientific disagreements, often valued by scientists as a demonstration of vitality and social responsibility within the scientific community, have weakened the public credibility of science. The role of scientists in controversies, for example, has raised questions about the distribution of access to expertise, and about the public accountability of scientists who appear to be associated with particular interests.

A third type of controversy over science develops when individual expectations conflict with social or community goals. Characteristically, such controversies—and they are ubiquitous—are framed in terms of "individual rights." If a water supply is fluoridated, universal vaccination required, or a course of study mandated in the public school curriculum, everyone must comply with the decision and share its consequences. If the sale of a product or drug is prohibited, those who want it are denied. Creationists see the teaching of evolution as a threat to their right to maintain the religious faith of their children, and this controversy over teaching evolution in public schools has persisted even after a United States Supreme Court decision seemed to bring closure to the issue (Nelkin 1984). Work in the neurosciences has in the past been controversial for its implications for behavior control (Shapiro 1974; Nelkin and Tancredi 1994). In this "decade of the brain" (as declared by the National Institutes of Health), we are likely to see further disputes in this area. Theories of sociobiology have provoked fears that biological determinism will serve to justify a reactionary politics of state control infringing on individual rights (Caplan 1978). The billions to be spent on the Human Genome Project exacerbate these fears. Some fears are legitimate, others less so; but they all play on existing tensions in U.S. society over the appropriate role of government and regulation, and the extent to which community values may intrude on the freedom of individuals (Nelkin and Lindee 1995).

There are other types of disputes without these overt moral dimensions. The growth of big science—the now-defunct Superconducting Supercollider (SSC), the space station Freedom, and the Human Genome Project—generated conflicts over questions of equity in the distribution of resources within science. The commercial appeal of biotechnology has generated disputes over patenting and property, as those concerned with technological innovation in a competitive market conflict with those who feel the public interest would be better served by allowing more open communication of new ideas. And incidents of scientific misconduct are the source of heated disputes over the accountability of science and its capacity for self-regulation. These policy issues mainly engage scientists and policy-makers, not the public (Dickson 1984). But they are inevitably debated in the media and so contribute to scarring the public face of science. And they add to the persistent concern that prevailing popular sentiments have little influence over science policy, that technical complexity, specialization, and corporate interests prevent meaningful public control.

These diverse controversies over science express two aspects of the public face of science that call for further analysis. First, they pose political challenges to science and science policy. Second, they indicate growing moral concerns about scientific practices.

Controversy as a Political Challenge

In many ways, controversies over science represent a decline in public trust. The growing importance of expertise in policy decisions seems to limit the democratic process (Goggin 1986).[3] Critics question the ability of representative institutions to serve the public interest (Ezrahi 1990). They are asking about research priorities: Is science for the public or simply for the advancement of scientific careers? Do the applications of science benefit society or do they simply fulfill narrow economic goals? In many protests, activists demand greater public involvement in decisions, though in fact only about 3 percent of the country's adults are both attentive to science policy issues and sufficiently literate scientifically to understand and assess the arguments underlying policy disputes (Miller 1990).

The nature of the political challenge varies with the issue and its constituency. Conflicts over science draw support from different publics. Some people have immediate stakes, as in the health risks or in the social and economic consequences of a scientific development. But many protests attract people with no direct interest at stake, and their opposition reflects a broad social, political or ideological agenda (Douglas and Wildavsky 1982; Downey 1986). In the debate over ozone depletion, for example, there is no natural constituency, because the only affected interests are future generations. And while some critics of biotechnology are concerned about the specific economic or environmental impacts of biotechnology applications, others have broader ideological concerns about the propriety of tampering with life.

Most activists in science-based policy disputes are middle class and educated people with sufficient economic security and political skill to participate in a social movement (McCarthy and Zald 1973). But participation in protests is not necessarily tied to traditional political alignments. For example, those involved in the animal rights and ecology movements identify themselves as liberals. But conservative right-to-life groups oppose fetal research, and fundamentalists, in the creation controversy, seek to block the teaching of scientific theories that offend their faith. Focused on particular problems, science controversies attract people whose concerns rest more on the nature of the issue than on their prior political orientation as liberal or conservative. What links these diverse groups is their demand for greater accountability and increased public control. Alain Touraine (1980) has described such disputes as a reaction against technocracy. While hardly a new theme, it is part of the significance of today's scientific disputes.

Controversy as a Moral Crusade

Political demands in science controversies are usually cast in the moral rhetoric of rights, a rhetoric rooted firmly in the political history of the United States (Jonsen 1991). The tendency to formulate problems in terms of distinct, overarching moral principles, apart from their social context, was nurtured by the religious tradition of Calvinism, and moralist thinking later permeated secular thought in the United States through the tradition of Puritanism (Miller 1962).

Today, this tendency is reflected in the revival of bioethics as an influential profession, in the discourse of social movements and their insistence on moral absolutes, and in the polarization of many disputes into questions of right and wrong. The increasing number of protests that are framed in terms of "rights" express a kind of moral fundamentalism. Animal advocates call for animal rights; antiabortionists make claims for fetal rights; scientists claim the right to conduct their research without unwarranted intervention; creationists claim their right to choose the theories taught to their children; and environmentalists advocate the rights of future generations.

Some claims to rights are based on obligations; rights claims may be a practical condition, necessary to fulfill certain tasks. Thus, government agencies claim the right to constrain individual freedom in order to carry out their mandated responsibilities. Other claims to rights are based on utilitarian arguments; certain rights are valued because they maximize the public good. Scientists, for example, argue that the acquisition of knowledge is so important for the long-term interests of society that freedom of inquiry must override other considerations. Some base their claims to rights on moral or religious premises; others on the libertarian assumption that individual autonomy is an ultimate value in itself. But whether justified in terms of natural rights, obligations, or traditions, rights claims become moral imperatives. They are considered nonnegotiable, leaving little room for compromise or accommodation, and their violation is thought to warrant extreme reaction.

In some controversies, of course, claims to rights confuse moral categories with strategic goals, and they are little more than ad hoc responses in competitive situations. Indeed, the rhetoric of rights may be simply a way to limit negotiation by elevating instrumental behavior to the level of a moral imperative. Rights claims may indeed be the central issue in a dispute, but they may also be merely a tactic, a way to gain public support in a controversial political context. In either case, those who make rights claims inevitably limit the rights of others, exacerbating controversy. Framing disputes in terms of fundamental value differences contributes to the difficulty of resolution.

Science as Strategy in Value Disputes

Further compounding the ideological complexity of science controversies is the strategic involvement of scientific expertise. In some disputes—over the diet-cancer connection, recombinant DNA, or ozone depletion—scientists initiated controversy by raising questions about potential risks in areas previously obscured from public knowledge. But scientific expertise has become a crucial political resource in all policy conflicts, even those, like surrogate motherhood, that involve complex social decisions but only limited technical sophistication. To all the protagonists in policy disputes, access to knowledge and the resulting ability to question the data used to legitimize decisions are an essential basis of power and influence (Benveniste 1972).

The authority of scientific expertise rests significantly on assumptions about scientific neutrality (Proctor 1991). The interpretations and predictions of scientists are judged to be rational and immune from political manipulation because they are based on data that has been gathered through objective procedures. But in scientific controversies expertise is enlisted both to defend and to challenge the legitimacy of policy decisions. Just as industry advocates use technical expertise to support their projects, so too do protest groups who oppose them. Among animal rights advocates are scientists who debunk the need for using animals in research, question its scientific justification, and lay out alternative research strategies. Even fundamentalists who seek to have the Biblical account of creation taught in the public schools present themselves as scientists and claim that creationism is a valid scientific theory.

As a result, although political and moral values motivate such disputes, the arguments often focus on technical questions.

Shifting difficult social dilemmas to the technical arena can be tactically effective. In all disputes there are broad areas of uncertainty open to conflicting scientific interpretation, and decisions are necessarily made in a context of limited knowledge. Thus, winning a dispute may hinge on the ability to manipulate knowledge and to challenge the evidence presented to support alternative policies. Expertise, in effect, is reduced to a weapon in the political arsenal of competing groups.

But as expertise becomes a resource exploited by all parties to justify particular moral and political claims, it becomes difficult to distinguish scientific facts from political values. Protagonists on both sides of a dispute may use the work of "their" experts in ways that reflect biased judgments about social priorities or about acceptable levels of risk. For this reason, debates among scientists have altered the public face of science, revealing how value premises can shape the data considered important, the alternatives weighed, and the issues regarded as appropriate (Hilgartner 1992). While the willingness of scientists to lend their expertise to various factions in publicized disputes is also a manifestation of social responsibility, it has undermined assumptions about the objectivity of science. And ironically these are the very assumptions that have given experts their authority as the neutral arbiters of truth. (See also Woolf, this volume.)

Beyond seeking technical resources, those engaged in controversies over science must organize their activities to broaden their political base. Many protest organizations—like animal rights and ecology groups—rely on the backing of a direct mail constituency to provide political and financial support for their causes. To do this successfully, they must generate dramatic, well-publicized events; capturing public attention and political interest requires engaging the media (Nelkin 1987). Moving beyond routine political activities like lobbying or intervention in public hearings, protest groups engage in highly visible actions—laboratory break-ins, street demonstrations, and civil disobedience. They involve colorful or visible writers and activists such as Jeremy Rifkin, Peter Singer, Ralph Nader, and Paul Brodeur, or film personalities and politicians, to generate media appeal. And they exploit visual images. The gruesome photographs projected by animal rightists attract broad support for this social movement. Pictures of third trimester fetuses, or the ultrasound images that seem to show the fetus waving an arm or sucking its thumb have encouraged concerns about fetal research and tissue transplants. Rhetorical strategies are also critical. To scientists engaged in fetal research, the fetus is a "tissue"; to opponents it is a "baby." These verbal and visual images are effective means to engage the broader public in policy disputes. They help to shape public perceptions of science. Protest actions in effect become a way to create an image of science—to configure, refigure, or disfigure its public face.

The Resolution of Conflict

Some people are more susceptible than others to such images of science, depending on their interests, experience, and values. Existing biases may assume far greater importance in shaping public perceptions than any details of scientific verification. To be sure, in disputes reflecting economic interests or concerns about risk, new information may change the character of controversy. The arguments over the relationship between diet and cancer and over the environmental effects of chlorofluorocarbons (CFCs) have shifted in response to changing data.

But the history of moral disputes offers little evidence that technical arguments change well-entrenched beliefs. In moral controversies, cosmetic changes are fundamentally ineffective in changing the public face of science. The wrinkles may be covered, but they do not disappear. If basic moral premises or ideological principles are at stake, compromises and accommodations have limited effect. For example, no amount of data could change the premises underlying the conflicts over fetal or animal experiments, disputes that have persisted despite changes in research practices and policies. In the creation controversy, too, efforts to convince believers that there was a scientific consensus about the principles of evolution failed to sway those morally committed to the cause.[4] The participants in such disputes base their positions on well-entrenched beliefs, and their conflicting visions preclude easy resolution.

Although such disputes are not easily resolved, they may come to temporary closure. A Supreme Court decision[5]—at least for a while—foreclosed further legislative efforts of creationists, though they continued with increased fervor to bring their demands to local school districts. In some cases powerful protest groups have been able to exercise sufficient political leverage to affect science: antiabortionists and animal rightists have had a striking influence on research practices, ranging from reforms to outright bans on certain types of research. Some controversies have resulted in the withdrawal of government funding from projects (such as the ten-year ban on fetal tissue research). Scientists hoping to avoid conflict have themselves moved away from certain areas of research (for example, on the genetic basis of criminality). One can only speculate on the influence of disputes on young people contemplating scientific careers.

Ultimately, the implementation of science policy depends on public acceptance—or, at the least, public indifference. Efforts to foster greater acceptance of science are numerous. Experiments in negotiation and mediation have proliferated as a means to resolve disputes (Susskind and Weinstein 1980). Participatory experiments have led to the appointment of citizens to the advisory committees and institutional review boards overseeing research. Peer review groups, consensus panels, and special commissions have been created as a means to build public trust (Jasanoff 1990). And there have been numerous efforts to work with the media to create a better public image.

While many of these efforts are intended to expand communication, controversies have also led to policies for the suppression of information that might arouse public concerns. In the late 1980s, the increasing governmental focus on international technological competition encouraged secrecy and reduced opportunities for participation. Access to information in some controversial areas became more and more difficult. After the Chernobyl accident, federal agencies issued gag orders to energy agency officials and the several thousand scientists at national laboratories. They feared that disclosure of information to the press would result in hasty and inappropriate public responses to the controversial U.S. nuclear power program (Nelkin 1989). But such inclinations towards secrecy tend to backfire because they generate a very hostile press.

In the end, few conflicts over science are ever really resolved. Even as specific debates seem to disappear, the same issues recur in other contexts. Expressing moral judgments and personal beliefs as well as economic and political interests, and driven by a rhetoric of good and of evil, of right and of wrong, many scientific controversies turn into nearly religious crusades.

Most cosmetic efforts at resolution are—like a face lift—only temporary; for they sidestep fundamental changes in the public face of science. These changes reflect the maturity of science and the growing awareness of the powerful influence of this large and costly social institution. They reflect the persistent efforts of many groups to reassess the moral values, social priorities and political relationships shaping public policies. And they reflect the tensions that are endemic in U.S. society in the 1990s—tensions that seem to be shattering the social consensus that has long supported science as well as other communal activities.

Notes

1. The cases described in this paper are partly drawn from Nelkin (1992).
2. This section draws on a number of books that have provided case studies and analyses of scientific controversies, including Holton and Morison (1979); Mazur (1981); Engelhardt and Caplan (1987); IOM (1991); and Nelkin (1992).
3. For a literature review, see Nelkin (1987).
4. Engelhardt and Caplan (1987) have included many cases demonstrating the difficulties of resolving disputes.
5. *Edwards v. Aguillard*, 107 S.Ct 2573 (1987).

Bibliography

Benveniste, G. 1972. *The Politics of Expertise.* Berkeley: Glendessary Press.

Brzezinski, Z. 1970. *Between Two Ages: America's Role in the Technetronic Era.* New York: Viking Press.

Caplan, A. L., ed. 1978. *The Sociobiology Debate.* New York: Harper.

Dickson, D. 1984. *The New Politics of Science.* New York: Pantheon.

Douglas, M. and Wildavsky, A. 1982. *Risk and Culture.* Berkeley: University of California Press.

Downey, G. 1986. Ideology and the clamshell identity. *Social Problems* 33:101–117.

Engelhardt, T., Jr. and Caplan, A. L. 1987. *Scientific Controversies.* New York: Cambridge University Press.

Ezrahi, Y. 1990. *The Descent of Icarus: Science and the Transformation of Contemporary Democracy.* Cambridge: Harvard University Press.

Fischer, F. 1990. *Technocracy and the Politics of Expertise.* Newbury Park, CA: Sage Publications.

Goggin, M., ed. 1986. *Governing Science and Technology in a Democracy.* Knoxville: University of Tennessee Press.

Hilgartner, S. 1992. Who speaks for science: Disputes among experts in the diet-cancer debate. In D. Nelkin, ed., *Controversy: The Politics of Technical Decisions.* 3rd ed. Newbury Park, CA: Sage Publications, 115–129.

Holton, G. and Morison, R., eds. 1979. *Limits of Scientific Inquiry.* New York: Norton.

Institute of Medicine (IOM). 1991. *Biomedical Politics.* Washington, DC: National Academy Press.

Jasanoff, S. 1990. *The Fifth Branch: Science Advisors as Policymakers.* Cambridge: Harvard University Press.

Jasper, J. and Nelkin, D. 1992. *The Animal Rights Crusade: The Growth Of A Moral Protest.* New York: Free Press.

Jonsen, A. R. 1991. American moralism and the origin of bioethics in the United States. *The Journal of Medicine and Philosophy* 16:113–130.

Krimsky, S. 1982. *Genetic Alchemy.* Cambridge: MIT Press.

Krimsky, S. 1991. *Biotechnics and Society.* New York: Praeger.

Maynard-Moody, S. 1992. The fetal research dispute. In D. Nelkin, ed., *Controversy: The Politics of Technical Decisions.* 3rd ed. Newbury Park, CA: Sage Publications, 3–25.

Mazur, A. 1981. *The Dynamics of Technical Controversy.* Washington, DC: Communication Press.

McCarthy, J. and Zald, M. 1973. *The Trend of Social Movements in America: Professionalization and Resource Mobilization.* Morristown, NJ: General Learning Press.

Miller, J. 1990. *The Public Understanding of Science and Technology in the United States, 1990.* Washington, DC: National Science Foundation.

Miller, P. 1962. *The New England Mind.* Cambridge: Harvard University Press.

National Science Board (NSB). 1989. *Science and Engineering Indicators, 1989.* 9th ed. NSB-89-1. Washington, DC: U.S. Government Printing Office.

Nelkin, D. 1984. *The Creation Controversy.* New York: Norton.

Nelkin, D. 1987. Science, technology and public policy. *Newsletter of the History of Science Society* (January).

Nelkin, D. 1989. Communicating technological risk. *Annual Review of Public Health* 10:95–113.

Nelkin, D. 1994. *Selling Science: How the Press Covers Science and Technology.* 2nd revised edition. New York: Freeman.

Nelkin, D., ed. 1992. *Controversy: The Politics of Technical Decisions,* 3rd ed. Newbury Park, CA: Sage Publications.

Nelkin, D., and Lindee, S. 1995. *Supergene: The Powers of DNA in American Culture.* New York: Freeman.

Nelkin, D., and Tancredi, L. 1994. *Dangerous Diagnostics.* Chicago: University of Chicago Press.

Proctor, R. 1991. *Value-Free Science?* Cambridge: Harvard University Press.

Richards, E. 1988. The politics of therapeutic intervention: The vitamin C and cancer controversy. *Social Studies of Science* 18:653–701.

Rosner, D., and Markowitz, G. 1991. *Deadly Dust: Silicosis and the Politics of Occupational Disease in 20th Century America.* Princeton: Princeton University Press.

Salomon, J. J. 1977. Crisis of science, crisis of society. *Science and Public Policy* (October): 414–433.

Shapiro, M. 1974. Legislating the control of behavior. *Southern California Law Review* 277:237–356.

Susskind, L., and Weinstein, A. 1980. Towards a theory of environmental dispute resolution. *Environmental Affairs* 9:311–356.

Touraine, A. 1980. *La Prophecie Anti-Nucleaire.* Paris: Edition du Seuil.

6

How Large an R&D Enterprise?

Daryl E. Chubin

A palpable malaise has overtaken the relationship between government and science. In this chapter, Daryl E. Chubin diagnoses two major manifestations of the malady: changes in the scientific enterprise from the time of its inauguration immediately following World War II; and a competition among different values of the research system. The first changes have given us a different science than existed nearly a half-century ago, although that new science is governed by the same institutions and same expectations. The competition has caused a subtle tilting of priorities in the research system and even a relative "deprivation" of some important values that has alienated research constituencies. Put simply, it is possible that the research system has grown beyond the ability of the government to support it in its entirety.

The research system will have to recognize some difficult truths in order to begin the road to recovery. The multiple roles of universities in education, research, and economic development can cause conflicts. Research funding has become big politics as it has become big money, and politics means controversy, compromise, and tradeoffs. Professional "science watchers" perform "institutionalized scrutiny" of the research system. Given this diagnosis, there is no magic bullet to cure what ails the government-science compact, no simple injection of money or labor or autonomy that will restore its prior health. Chubin's diagnosis instead means that systemic changes must be made, and he points to some leaders within the science policy community who are in fact engaging in "systems" and "ecological" thinking. The new compact between government and science must also recognize that the partners are not as different as they thought when the original compact was drawn up: both politics and science are part of the same national environment and geared toward the same common goals.

Chubin prescribes four regimens to restore the vitality of the government-science compact that can all be characterized as strengthening the links between society and science. By placing science in its environmental context—society—we may create a new compact for science that is less fragile, more productive, and even more democratic. [Eds.]

Introduction

The dilemmas facing U.S. science today and for the forseeable future can be distilled in a single question: How can we preserve what have been the strengths of the "federal research system" while adapting to changes in the system itself? Size, diversity, competition, and productivity are at once assets and burdens.

"How large should the research and development (R&D) enterprise be?" raises another question, posed in a 1991 report of the congressional Office of Technology Assessment, *Federally Funded Research: Decisions for a Decade:* "How much is enough?" Questions of magnitude or scope pit expectations against historical realities, and spur yet other questions.[1] To wit, how much for what? If the 5 percent of the total federal budget that currently goes to R&D is too little (OTA 1991: 11), then would 10 percent be too much or just right? We have few metrics to help us decide, but most scientists and "science watchers" harbor lofty aspirations and even higher hopes.[2]

This chapter approaches the "how much?" question from various angles or sources of authority—professional, institutional, political, and empirical. It begins with the issues of the 1990s that the OTA report associated with "science policy." Like all issues, they evolve yet retain a perennial quality, challenging experts inside and outside government to find new ways of addressing them. The bigger challenge may be finding new ways of thinking about them. The chapter concludes, therefore, with thoughts on "big thinking"—why more of it could render science policy far less consequential than it seems to be now.

How large an R&D enterprise is not necessarily the wrong question; but it is far less important than wondering who is asked to respond to the question and how institutions concerned with the welfare of R&D in turn react to these responses.[3] I think of this, and public policy making in general, as the process of participatory politics. Who has the power to speak and be heard—and to get others to act in bold and perhaps unpredictable ways?

Diagnosing the Problem

The hopes and aspirations of the R&D enterprise can be traced to the post–World War II compact between the federal government and particularly academic science.[4] The familiar benchmark for the compact is Vannevar Bush's *Science: The Endless Frontier*, which gave rise to expectations that feed the current dilemma.[5]

Over a quarter-century ago, however, we were reminded that there are limits to growth: "We simply cannot expect the extraordinary rate of growth in Federal science expenditures . . . to continue unabated. . . . Since exponential growth is proportional to what exists, it competes with itself in a race, requiring ever more effort just to maintain momentum (Sanders and Brown 1966: 93–94). Recently, Erich Bloch, the former director of the National Science Foundation (NSF), has called the current situation "a mid-life crisis of a system that has had significant growth and, because of this growth, has developed fissures in its structure and cultivated administrative overreach. The academic research system is now facing a plateau, a levelling off, and is having difficulties coming to grips with its unfulfilled aspirations" (1991: 147–148).

So what is to be done, and by whom, when the gap between aspirations and opportunities yawns so widely? A return to fundamental questions has obvious appeal. The OTA report asks what scientific and engineering research in the United States should attempt to accomplish:

> 1. Is the primary goal of the Federal research system to fund the projects of all deserving investigators . . . ? If so, then there will always be a call for more money, because research opportunities will always outstrip the capacity to pursue them.
>
> 2. Is it to educate the research work force, or the larger science and engineering work force . . . ? If so, . . . then . . . preparing students throughout the educational pipeline will assure an adequate supply and diversity of talent.
>
> 3. Is it to promote economic activity and build research capacity throughout the United States economy . . . ? If so, then the support should be targeted . . . to pursue applied research, development, and technology transfer.

> 4. Is it all of the above and other goals besides? If so, then some combination of these needs must be considered in allocating Federal support.
>
> Indicators of stress and competition in the research system do not address the question of whether science needs more funding to do more *science.* Rather, they speak to the organization and processes of science . . . on which the system is built and that sustains its vigor. (1991: 12)

With President Clinton's deficit-reducing budget, the R&D agencies generating strategic plans, research communities engaging in priority-setting exercises (many under the auspices of the National Academy complex), and inquiries into universities' indirect cost recovery practices for federally funded research, conjectures about the size and goals of the research enterprise—at least the federal portion—continue to mount. The conjectures—data and diagnoses—need to be put into some perspective.

Then and Now

Since 1966, the approximate inflection point in the geometric growth of science funding, it has appeared inevitable that leveling would occur (Price 1961; Cozzens et al. 1990). The research community of the 1950s and 1960s enjoyed a funding situation very different from the situation today. Then the research environment was characterized by: (1) fewer researchers, ample job opportunities, and a more homogeneous work force (white males), and this composition was reflected in the student population; (2) fewer research universities (the principal performer) with a greater concentration of federal resources (the principal sponsor); (3) little international competition or concern about U.S. research performance—the United States was clearly dominant; (4) a compact that entrusted the judgment of scientific merit to expert peers in return for scientific progress that would benefit the nation's national security, productivity, and quality of life; and (5) researchers' expectations of sustained federal funding, now pejoratively called an "entitlement" mentality among scientists.

That was then and this is now. I would suggest that when funding falls short of expectations, when the number of deserving researchers is such to deprive some of the chance to pursue promising opportunities (or pursue them as fast or as fully as one hoped), the result is a relative, not an absolute, deprivation. "Relative deprivation" is a social science concept that highlights the disjuncture between federal funding trends on the one hand, and institutional and personal angst on the other. This angst is what the panoply of speeches, surveys, and reports all capture. Does knowledge of this paradoxical disjuncture lessen the pain or lift the morale? Not necessarily. But there is an implicit, corollary question: Does the research community expect that the federal government will fund every deserving researcher? It cannot. Some adjustment in thinking in the research community—that is, in the government-university compact—is needed. As Robert Rosenzweig, then president of the Association of American Universities (AAU), has noted: "University people are talking a little differently from the way they used to talk. They're questioning whether the stability and solidity of the relationship—things that were taken for granted in the past—are less likely to be taken for granted now. All that is healthy. I thought for a long time that university people had quite unrealistic expectations about governmental behavior."[6]

From the federal perspective, research universities and their faculties are being unrealistic. To universities, however, the cost of research is outpacing the ability to raise money.[7] Research universities' struggle to accommodate their many missions may itself feed the malaise afflicting researchers (GUIRR 1992).[8]

Roland Schmitt, president of Rensselaer Polytechnic Institute, has brought the institutional-interpersonal disjuncture into sharp relief. Schmitt argues that

> during the 80s there was significant, real growth in the support of academic research; it became broader (more institutions), deeper (more people); less tied to faculty positions; and less dependent on Federal support.
>
> As a result of these changes . . . [there has been] a disturbing erosion of the sense of citizenship and collegial behavior in academe; in its place has arisen an "academic entrepreneurship" of individuals housed under the institutional umbrella. Universities have become holding companies for these research entrepreneurs. . . . Institutional loyalty has given way to outside loyalties: [to] sponsors and other external agents. New, young investigators are thrown into this maelstrom early in their careers, often without the protective cover of a faculty position and certainly without tenure. Many are ill-equipped to deal with these circumstances so early in their careers. (1992: 2)

Deprivation Caused by "Competing Goods"

Schmitt is correct, but the malaise is as strong among senior investigators as among their younger colleagues with whom, in effect, they are competing for scarce dollars—set-aside programs notwithstanding. This competition is just one of the dilemmas facing the federal research system of the 1990s, as OTA has pointed out. Table 6.1 shows the tensions that require choices between "competing goods." All cannot be funded or otherwise supported. These dilemmas confront not only Congress, but all participants in the system. Their choices, and the negotiations over them, will determine the size and shape of the research enterprise. The ways those choices are made and the voices paid heed in the negotiations will reinvent the compact.

Schmitt suggests that the new compact between the federal government and the academic research community "must begin with an understanding of why public funds are used to support research." Then academic scientists and politicians and business leaders "must reconcile their respective priorities in establishing the new compact." At a minimum, says Schmitt, it must encompass the following:

> The academic research establishment itself must explicitly embrace the need to link its research to both education and utility. . . . It cannot claim its domain to be totally autonomous, driven only and solely by intrinsic values that are not shared by the public at large.
>
> Federal agencies . . . must also embrace this linkage. The ideological arguments over whether this linkage constitutes "industrial policy" or "technology policy" should be declared moot!
>
> Government funding agencies as well as private industry and philanthropists ought once again . . . give high priority to building world-class institutions and departments. . . . [W]e have neglected the very homes in which our researchers work, interact, and educate. (1992: 3–4)

Schmitt is prescribing institutional remedies and calling for strategic planning at a time when the public image of science and particularly of universities has come under a cloud. Indirect cost recovery, misconduct in research, and academic earmarking are all symptoms of a government-science (not just university research) contract that needs to be revised. We are living that revision. It will

Table 6.1 Tensions underlying policy choices in the federal research system

centralization of research planning	*and*	pluralistic, decentralized agencies
set-aside programs	*and*	mainstreaming criteria in addition to scientific merit and program relevance (e.g., ethnicity and gender)
peer review-based allocation	*and*	other funding decision mechanisms (agency manager discretion, congressional earmarking)
concentrated excellence	*and*	regional and institutional development
dollars for facilities	*and*	dollars for research projects
large-scale, multiyear capital-intensive, high-cost per investigator initiatives	*and*	individual-investigator and small-team, 1–5 year projects
training researchers in the model of their mentors (and creating more competition for funds)	*and*	encouraging a diversity of career paths (and easing competition for funds)

Source: Adapted from OTA (1991, table 1-1).

strengthen all partners—eventually. For now, how might we proceed?

A Prelude to Prescriptions

A memorable line from the 1970s warns of the task ahead: "Do not adjust your mind—there is a flaw in reality." Diagnosing the issues that have been prominent features of the contract reminds us that this is not about just the federal government and universities. Today, business and industry are significant research patrons, at least equal in their support of science and technology with the government. Their role as partners, as well as that of the states, in abetting technology transfer across national and sectoral boundaries in the global economy, complicates what is meant by "science policy." How this policy should be connected to and demarcated from "technology," "industrial," "trade," "immigration," and "education" policies will persist as a focus of debate. That is as it should be: If we seek to understand and intervene in the research system, we need to discern its working parts. Put another way, changes in so-called exogenous variables may be more significant than the institutional spheres under examination. It is the relationships that matter.

I cannot do justice here to these relationships. In the hope of enlarging the canvas, however, I describe three clusters of issues that must be recognized, if not addressed, as part of the reinvention of the government-university compact. These issues all impinge upon the relationships between the institutional spheres. They are: the multiple missions of universities, political dilemmas, and professionalized scrutiny of R&D.

Multiple Missions of Universities

The missions of universities in the 1990s are undergraduate education, graduate training, and intellectual property. Administrative efforts in education center on changing demographics, and especially the recruitment and retention of women and U.S. minorities to undergraduate study and specifically to majors in science and engineering fields. Issues of finance, classroom culture, and institutional commitment to student achievement mingle with faculty and governance issues.[9] What is the proper balance between research and undergraduate teaching? More directly, how can the rewards for excellence in teaching and advising (that is, for interventions both in and outside of the classroom) be structurally improved?[10]

With respect to graduate training, universities and departments must consider the preparation of Ph.D.s for careers beyond academic research. The model whereby a productive mentor would reproduce himself ten, twenty, or thirty times over may be dysfunctional for the 1990s and beyond if the career path of the new Ph.D. is intended to duplicate that of the mentor. This is not the same as saying there are too many researchers. It may mean there are too many academic researchers, or researchers at Ph.D.-granting institutions, which is the cadre that relies on the federal government for research support. The pattern of federal graduate student support, seen in table 6.2, is unmistakable: a shift away from fellowships and traineeships and toward research assistantships. If research grants had not burgeoned over the past 20 years, such support would not exist. Now the recipients—an expanding pool of competent researchers—are full-fledged competitors. Meanwhile, there are too many other national missions that must be served.[11]

Finally, universities are creators and retailers of intellectual property. Research productivity has many faces—basic knowledge, technological innovation and transfer, and the expertise embodied by faculty and other professional staff. This makes the university an agent of economic development, both as a resource and a catalyst.[12] It has attracted financial roles and responsibilities and fostered an entrepreneurial spirit, which some claim is a distraction from its original educational mission and a compromise of values; others hail it as proof that the university has become a responsive institution of the nation as well as the local community. The "knowledge-plus" era of universities—a quiet reality for most of this century (Geiger 1988)—is now upon us full-blown.

Political Dilemmas

Chief among the exogenous variables that affect the fortunes of universities are the perceptions and actions of policy makers, particularly Congress. These dilemmas center on the place of R&D in the discretionary budget. But the roots of the dilemma ensnare the political culture of science advising, executive-legislative gamesmanship, and the congressional budget process.

First is the conflict, underappreciated by those outside Washington, between authorization and appropriations subcommittees. The former provide oversight to the research agencies and their programs. But their will is increasingly thwarted by appropriations subcommittees faced with impossible tradeoffs, such as housing vs. particle physics, in the bigger budgetary picture. In the words of the chairman of the House Committee on Science, Space, and Tech-

Table 6.2 Federal support of science and engineering graduate students, 1969 and 1988, by type of support

	1969	1988
Fellowships and traineeships	56%	24%
Research assistantships	38%	66%
Teaching assistantships and other	6%	10%
Total students supported	51,620	54,852

Fellowships and traineeships were not reported separately in 1969, and they have been combined for presentation for the 1988 data.
Source: OTA (1991), figure 7-4, from NSB (1989), appendix table 2-18, and NSF (1970), table C-11a.

nology: "Scientists and politicians need to raise the level of debate about funding for science programs. They must move beyond the absurdity of trading these programs against each other. Cutting science budgets to pay for low-rent public housing or the job corps is a trap that ensnares the nation in deeper problems" (Brown 1992: 19).[13]

A related conflict is the growth, and growing contentiousness, of academic earmarking. As universities turn to lobbyists and appeal directly to distributive politics, some perceive the undermining of merit. Others perceive the building of a research capacity in regions, states, and institutions with a lesser tradition or a smaller concentration of research performers as both wise and egalitarian. OTA found that both of these positions lack empirical support: there are no output data on either the relative merit of proposals or the quality of science conducted at centers funded through earmarking, and yet the practice hardly redresses inequities (1991: ch. 3). It is not even clear that peer review and earmarking are directly competitive, though agencies find the latter practice disruptive of their planning.[14]

A third political dilemma is fraught with perceptions run amuck: the problem of indirect cost recovery for research. Like earmarking, it is vulnerable to the hyperbole of front-page news and deserves a separate, in-depth analysis (see Likins and Teich, this volume). Binary oppositions vie for attention: universities versus the federal government; academic administrators vs. faculty researchers; the Office of Management and Budget vs. congressional committees; and federal auditing agencies vs. federal grant-giving agencies.

At the center of the controversy is the doctrine of "full cost recovery" and arcane and idiosyncratic university accounting procedures. While one can count what the federal government spends on extramural research, it is much harder to know how much it costs (OTA 1991: ch. 6). The National Institutes of Health (NIH) is trying to find out, and universities will slowly (and apparently with reluctance) disclose what is covered under indirect cost. The appearance of impropriety—despite university protests that the rules (or their enforcement) have changed—will heighten the demand for accountability and undermine the very logic and efficiency of cost pooling. This will tax universities and divert resources from the performance of research, but to regain the public trust the sacrifice must be made.

A final political dilemma is the call for explicit rationales in support of R&D: priority-setting and criteria of choice. The concerns are identical to those of Alvin Weinberg (1966) more than a quarter-century ago. Today, the budget deficit looms over the debates championing "R" vs. "D," "big" vs. "little" science, young vs. old investigators, etc. Often, these debates are cast as the present against the future, scientific merit against national need. And coincidentally or not, all the R&D agencies (and even a few fields, such as astronomy, ecology, and space science), are now engaged in high-profile strategic planning. Whether this merely changes the rhetoric surrounding the process or actually results in different choices remains to be seen. In the absence of sound measures of returns on investment and a federal policy-making apparatus that precludes cross-cutting analysis of the whole R&D portfolio, it is unlikely anything will soon change. But the tradeoffs and politics are becoming brutally clear.[15]

Professionalized Scrutiny of R&D

The cast of characters in the policy-making process has changed. The post–World War II generation of science policy advisors transferred their experience and halos as scientists (mostly physicists) into the political arena. They were "cold warriors," many of them participants in or direct descendants of the Manhattan Project.[16] This is not to imply a single-mindedness, but instead an approach to problem-solving shaped by living the transition from wartime to peacetime. It was an era for celebrating the fruits of democracy (Mannheim 1952; Chubin 1985). An attenuation of the influence of physicists in policy making has been openly suggested; some think that biologists will succeed them (Kantrowitz 1992).[17]

I am hopeful for more diverse participation that will change the character of the process. A new generation of professional policy analysts, some with graduate degrees in policy and social science, is succeeding the science policy architects and advisors. Some migrate to policy careers. Their expertise, in short, is different from their predecessors' (Chubin 1992a). To some it signals less authority for science in politics. It certainly suggests that the character of science policy will change in the current era. This has already occurred; professionalized scrutiny has been institutionalized in the federal R&D apparatus at the Office of Science and Technology Policy (OSTP), NSF, NIH, and the congressional support agencies (notably OTA and the Congressional Research Service [CRS]). In addition, the nonprofit stakeholders, including the National Academy complex, omnipresent think tanks and foundations, and oddities such as the Carnegie Commission on Science, Technology, and Government and the Howard Hughes Medical Institute, constitute a reservoir of analysis and advice for policy makers.

This institutionalized scrutiny has "upped the ante" for information, and underscores the potential role for data in policy making. It has also shifted the discourse away from the authority of credentialed scientists—many honored with Nobel prizes and other trappings of impeccable judgment—and toward the data and lack thereof about the federal R&D system. This can expose scientists as partisans who favor their subjective experiences in the research trenches against national trend data (Chubin 1987). And there's the rub: it is the "outsider" policy analysts who act "scientifically" in scrutinizing the scientists' support system, whereas the scientists advocate increased federal support based on their privileged "insider" status as participants in a research community. Such an appeal derives from the expectations of the longstanding government-science compact.[18]

The clash of generations is always predictable; it is the manifestations of restiveness and disagreement over what to do about it that are uncertain (and in retrospect what distinguishes one era from another). In an era of "competitiveness," the necessity of international cooperation in U.S.-initiated "megaprojects" becomes clear not due to cost alone. The nation's reputation as a "reliable partner" is also at stake. What is troubling is that "science" projects can be held hostage to a panoply of agreements that have consequences far beyond science. This is how national science policy is embedded in the international politics of the global marketplace.

Science policy today blurs the distinction between "R" and "D," manufacturing and trade, "us" and "them." Fundamental values associated with research take on the colorations of the agencies that support it and the missions redefined to justify its support. For example, former NSF director Walter Massey declared education, research, and competitiveness as interwoven missions for the agency. He spoke of "mission motivated" research, which emphasizes as a criterion in funding decisions "program relevance." This from an agency virtually synonymous with "peer review," which is still the cornerstone of decision making but under a broader definition, namely, "merit review."[19]

Another consequence of professionalized scrutiny of R&D is that it tends to insulate policy making from grass-roots influences and vests it in a multitude of federal and nongovernmental organizations that function as a kind of "knowledge elite." Regardless of one's conviction that spokespersons of research communities or policy analysts either confirm or disrupt the biases of policy makers, it is fairly obvious that ordinary citizens have a difficult time participating in this democracy of credentialed expertise. To make matters worse, science "popularizers" are disparaged and science journalists reviled by researchers who nonetheless continue to lament an elusive "public understanding" of science. In the era of a new compact for science, these unusual "boundary-spanners" will play larger roles as both watchers and critics of science (Goldsmith 1986; Chubin et al. 1987).

To recapitulate, money alone will not solve the problems of the federal research system. The federal government, the executive and legislative branches alike, is not organized to monitor and manage the pluralistic, decentralized R&D enterprise. If the politicians cannot link R&D funding to national goals, they will simply not look after the investments in a coordinated and flexible manner. The portfolio will continue to resemble an ad hoc collection of pet programs and projects that muster key support from key participants at propitious points in the policy-making process. The nation deserves better—a new, reinvented government-university compact.

Changing Roles: Rethinking for New Times

If it is likely that the government-university compact of the past forty-five years must be amended, then the issues are not simply ones of size, scope, and goals (Browning 1991). More important are the relationships that define the character of participation and the expectations of all concerned. For example, expectations for the careers of science and engineering Ph.D.s may need to change. The question may not be how much or how many, but what will these people do? How will they help us achieve a range of national goals? How can federal policy assist in this process?

Such questions point to this reciprocity: the need for new roles—and new understandings—that bind the federal government and universities in "common cause." The agencies may need to clarify their missions instead of taking on new ones, and acknowledge that resources will force choices. Taken together, such tradeoffs would give the federal government a more specialized posture toward R&D.

This posture will either reward research universities by concentrating R&D funding in the top ten, twenty, or fifty that historically have produced the most research and trained the lion's share of Ph.D.s (and therefore constitute a national treasure), or distribute federal resources more widely and according to criteria that augment scientific merit. Because science and engineering are capital-, instrument-, and labor-intensive pursuits, they are elitist. Not every institution can provide the facilities and expertise to do cutting-edge research. Yet the research university model offers that prospect and distracts universities from their other missions of education and utility.[20]

The federal government could be boldly interventionist or laissez-faire in its support patterns. It could champion strategic planning and priority-setting, both for itself and for the institutions that perform research.[21] For every research dollar awarded, more strings (in the form of demands to serve various national purposes simultaneously) could be attached. Yes, this might mean greater accountability, but it would recognize those institutions that are explicitly goal-driven.[22]

Universities likewise have several choices to make. If they find the federal government an overbearing patron, they could continue to wean themselves from federal research funding. However, most research universities' operating budgets are so dependent on federal monies that weaning is unlikely.[23] "Leveraging" federal monies with state, private, philanthropic, and international contributions is routine, but the strings that these patrons attach may be even more restrictive, short-term, and anything but value-neutral.

The current reward structure allows those who excel at entrepreneurial activity to buy out of undergraduate education. Will the university rediscover the student "client" and promote true educators whose scholarship and service in the classroom are deemed vital instead of inconsequential or diversionary (Boyer 1990)? Very few current or former university presidents and chancellors—Frank Rhodes, Donald Kennedy, Derek Bok, James Duderstadt, Donald Langenberg, Roland Schmitt, and Harold Shapiro may be the exceptions—have asserted the need to reinvent the university quite apart from reconstructing its ties to government and other social institutions.[24]

Compounding this reticence to lead is the wholesale turnover of science leadership in the professional apparatus. National Academy of Sciences (NAS) president Frank Press and AAU president Robert Rosenzweig have both become private citizens again. The election of Bill Clinton led to new appointments at OSTP, NSF, and NIH.[25] Turnover will provide opportunities, but as always, will create some discontinuity in the credibility of proven, familiar leaders to bridge their constituents' concerns to policy makers and the public.

Will new thinking about universities, their connection to national goal-attainment, and the catalytic role of the federal government come with a turnover in leadership? Is a wholesale recasting of university missions needed, or perhaps a more abrupt transition in what the government is willing to pay for?

The Enterprise as System

The most thoroughgoing challenges to conventional thinking about the government-university compact and science policy more broadly have come from a "systems" thinker with experience in industry and government who now directs an academic public policy program, and a chemist-cum-university vice president for research who is now a college president. The former is Lewis Branscomb, the latter Linda Wilson.

Here, in capsule form, is the crux of their challenges. In a 1989 hearing held by the Senate Commerce, Science and Transportation Committee, Branscomb asserted that:

> A lot of the way human society has used science and technology really derives from a sort of tradition of linear Newtonian mechanical thinking. [S]ome new ideas . . . come straight out of the study of ecology. . . .
>
> Ecological thinking is different from linear thinking. . . . And in the past whenever we have approached the general subject of science and technology policy, we've done so within the same intellectual framework that produced the scientific revolution [that led to] many of the problems that we face today.
>
> I have also for years thought that our policy for research is impeded by lack of an adequate language for describing how the different parts of the research system are going to benefit society. If we could get our way of budgeting R&D in the government defined in terms of principles that express where the benefits are going to flow, then we would redistribute our emphasis.[26]

To social scientists, Branscomb's thinking underscores the historical compulsion to consult scientists about science policy and ignore specialists who examine the structure of social institutions with which science—and policies for promoting and governing it—must interact.

This is where Wilson comes in. Though not a social scientist, her sensibilities reflect acute ecological thinking that unites the big structural picture with local, interpersonal, and institutional needs.

Wilson states that the U.S. research university enterprise

> was designed when most of the players were males, sole wage earners living near their workplaces, and supported physically, emotionally, and socially by their families. . . .
>
> The design features of our academic research enterprise—from tenure to research-support mechanisms, mentoring, and information-exchange strategies—were built on assumption of a context that permitted single-minded devotion to task, rather than a context for persons with multiple competing responsibilities spanning work and home.
>
> What has changed is the diversity of the roles of the players in the enterprise itself—where faculty are educators, researchers, entrepreneurs, policy advisors, peer reviewers, public relations managers, financial managers, and personnel managers—and in the diversity of roles played outside the enterprise. . . .
>
> We need to tap the entire pool of talent to strengthen, replenish, and renew our science and engineering work force. But the new "immigrants" to the science and engineering work force, women and minorities, bring some differences in expectations, some of which are based on care-giving responsibilities that have been traditionally assigned in a differentiated way. Redistribution of roles among men and women will take place, but the cultural roots are very deep and the stakes are very high. . . .
>
> Our near-term decisions therefore will involve personal attitudinal change, organizational change, and systemic change. These decisions themselves are interdependent—a set of human resources issues of a scope larger than we have had to address for many years.[27]

Wilson's remarks, delivered at the NAS in December 1991, aroused considerable ire. Some accused her of devaluing the competitiveness of research, which in turn is equated with diluting quality as detected, sorted, and rewarded through peer review-based decision making. Wilson never proposed such a compromise. Rather, some of the problems facing science "may be linked to the existing ground rules of competition."[28] "Macho" or "killer" science belongs to an earlier era—when men were men, scientists were men, and science was all about the mastery of nature instead of harmony among parts of the natural universe—in short, the difference between Branscomb's linear and ecological thinking.

Wilson's emphases, in other words, are not just "political." The way we talk reflects the way we think, and often shows how we are inclined to act. Holding the root values of the research university enterprise up for inspection allows for its entanglements—planned and unplanned—to be reconsidered. They are not necessarily wrong, but may be maladaptive for the times, offering models for behavior to a once-homogeneous research community that has become demographically, ethnically, and intellectually as heterogeneous as any of us might have ever imagined was possible. We should glory in it and exploit it, not ignore or deny it.

To treat the U.S. government-science compact as the model for all time (and indeed, all cultures) would continue to stigmatize changes as "problems," "anomalies," and departures from the "good ol' days." As the context has changed, so has the doing of science and the scientists themselves. The idea of federally funded research is still sound; its expression—the terms of discourse and the expectations of participants and interested observers—is like speaking in tongues. Shared meanings of such crucial concepts as "partnership" are lost, motives are doubted, and disappointments abound.

Prospects for a New Compact

To write about R&D as a system is to sound alternately preachy, self-righteous, and cynical—or at least disgusted with the system as it is and its portrayal by its dominant participants. This is a time of self-doubt and painful self-examination. It challenges scientists weaned on one system to contemplate another—one that is leaner, practical, efficient, fair, and yet uncompromising in its recognition and support of bedrock values: creativity, quality, and risk-taking. At stake, according to researchers, is the integrity of the research process and a unique national resource known as the university. Politics is viewed as inimical to researchers' interests, with the participation of nonspecialists in either the direction or conduct of their research an unnecessary evil. The dichotomy, of course, is false. That is what is meant by a new government-science compact. The need is for new habits of mind: new expectations about federal research funding; new understanding of the word "discretionary"; and a new responsiveness to questions raised by political actors, as coached by their support agencies and policy analysts.

The 1980s, cited as a watershed for R&D funding, also exposed a growing sophistication—or disdain, depending on your viewpoint—

about scientists as "just another interest group." It's payback time. The 1990s must demonstrate that the self-interest of scientists is in the best interest of the nation, if not the world. Lurking behind all this is what is quaintly called "the federal role." How should it change?

As we contemplate the size and purpose of the U.S. R&D enterprise, the new compact must:

1. *Bring people in.* The link between fiscal resources and human resources cannot be an afterthought. This is a time to be inclusive, not exclusive. The proliferation of set-aside programs at NSF stigmatizes both the recipients of these competitive, merit-based funds and the rationale for awarding them. If these programs are seen only as adjuncts to the core research funding in disciplines, then they will drive the wedge deeper between national goals and the utilization of talent. The winning strategy for the 1990s, as Linda Wilson exhorted, is one of participation—and federal agencies must spearhead such action.

Expanding the educational pipeline will be wasted effort, though, if career opportunities do not exist or access to them is not vigilantly protected. What is too often overlooked is an expanded role for scientists outside of academe and research. The future U.S. workforce will depend on the utilization of versatile, talented people whose scientific training gives them an added dimension. Doctoral education must encourage such versatility so that other professions and sectors of the U.S. social fabric—law, business, politics, education—may become more receptive and poised to hire and value Ph.D.s.

Finally, not only are women, U.S. minorities, and the physically disabled underrepresented in the traditional scientific careers, but they are also underparticipating in the networks of high-level decision making for the enterprise. Bernadine Healy of NIH, Walter Massey of NSF, Mary Good of the President's Council of Advisors on Science and Technology (PCAST), and Hanna Gray of the University of Chicago are anomalies by virtue of the status and visibility they have achieved. Women in science advisory circles are particularly scarce.[29] Full participation will require more than adherence to the letter of the law; it will mean fundamental change to the culture of science.

2. *Disabuse scientists and their patrons that the university can be the panacea for all the nation's ills.* As Paul Gray has put it:

> In the triad of universities, science, and government, it is not enough to ask for a rewritten contract which places a greater share on the universities. . . . [I]n the future less research will take place in a university setting. . . . Yet I think the role of the university will become, paradoxically, more rather than less important for the total enterprise. . . . I do think that universities will have to become more differentiated among themselves and that there will be a greater division of labor and of special areas of strength among them. (Abelson 1992: 9)

The universities are in the midst of a "shakeout." This is the legacy of the early 1970s when U.S. higher education was "democratized" in part due to the ample supply of Ph.D.s that could not be absorbed by the research universities. Today's shakeout will determine which models will dominate and which configuration of values will prevail. Some institutions will strengthen their toeholds on federal and industrial research resources and remain "research universities." Others will specialize in undergraduate education or fiercely embrace select fields of research, development, and graduate training. Some will become centers for statewide and regional technology innovation and economic development. All will rely on sizable and sustained support from multiple sources, each with its own agenda and influence over administrations, faculty, and students (that is, future workers) (Peters and Etzkowitz 1990; Barinaga 1991; Feller 1992).

Universities will have to decide how best to contribute to national goals—which functions to surrender and which to consolidate. No longer will a single model of competitive research blind universities to the other roles they can perform, and assuredly are playing.

3. *Think globally; connect research with development, and science with technology.* All aspects of the research system must be connected with other federal policies. The State Department may be more vital to U.S. R&D prospects in the next ten to twenty years than NSF or any of the mission agencies (Carnegie Commission 1992). How we collaborate formally through bilateral and multilateral agreements with other national governments, and establish cooperative programs that exchange personnel and share intellectual property will have geopolitical consequences far beyond the ordinary bounds

of science and technology. A united western Europe, an open market-oriented eastern Europe, and a fast-striding Pacific Rim have shifted both the terms of reference and the stakes. Globalization has changed the rules, the game, and the players (Schott 1991).

Because science and technology will not be university-centered, greater appreciation of the melange of sponsors and performers of research—especially the federal laboratories and agency intramural programs—will foster a more balanced policy (Marshall 1991; Sweet 1992). And universities will become more global entities, tied electronically and through international collaborations, than national resources serving single governments and "competitiveness" missions (Ausubel 1992). What is economically competitive will have to enrich many sectors and markets.

To go further, science policy—a small, esoteric specialty—may be an anachronism. Its practitioners need to identify with prominent domestic and international issues and "think bigger." Integrating science and technology policy to achieve national goals is something that policy analysts, comfortable in their professional community niches, must learn to do, and even transcend.

4. *Nurture science watchers and critics.* Arguably and ironically, the cause of public understanding, or at least appreciation, of science has been advanced most strikingly through the scandals and controversies—creationism, cold fusion, animal rights, fraud in research—portrayed on the front pages of our daily newspapers. While negative in content, these reports demystify science and humanize the people and institutions that are making news. This, too, is a form of "popularization"—displaying both the integrity and the mischief which a public must observe and criticize to get the whole picture of achievement and accountability in a participatory democracy.

My bias is that science watchers and critics will be social scientists—by credential or experience (science journalists exemplify the latter). They have a knack and seasoned eye for translating science and technology. Science watchers have a trained capacity for seeing science and technology in society, and can relate their interactions to social and economic progress (Ziman 1991; Newby 1992).

What this science watcher sees is that the size of the R&D enterprise will increasingly depend on how much we know about

R&D investments, how they perform, and how well we manage the portfolio, as much as on the percentage increase relative to other federal expenditures in any given year. But will the scientific community commit to reinventing what it has long taken for granted? How inclined will scientists be to enlarge their concept of citizenship, play more active roles in the political system, and forge a compact with policy makers and the public that is clear and bold?

How large an R&D enterprise depends on a vision incorporating ecological, systemic thinking. Neither a breadth of budget (for none will be robust enough) nor the benevolence of an earlier era conjured by the name of Vannevar Bush will rescue U.S. R&D. It is up to all of us.

Notes

1. For example, see C. Cordes, "Policy experts ask a heretical question: Has academic science grown too big?" *The Chronicle of Higher Education* (November 21, 1990):A1, A22.

2. The concept of "science watcher" was introduced by Carey (1988). This chapter is written in the spirit of science watching, as is discussed below.

3. This point is illustrated by two *Washington Post* editorials, "Paying for science," (March 7, 1992):A22; and "Science workers, science labs," (April 11, 1992):A24.

4. Parts of this section are based on Chubin (1992b).

5. Issued by the U.S. Office of Scientific Research and Development as "A Report to the President on a Program for Postwar Scientific Research" in July 1945, *Science: The Endless Frontier* has been reissued multiple times by the National Science Foundation, which the report helped create. Equally important but historically overlooked was the management-oriented *Science and Public Policy*, also known as the Steelman report for its author John R. Steelman, issued in 1947. See also Shapley and Roy (1985), U.S. Congress (1986), and Smith (1990).

6. See D. S. Greenberg, "A new nastiness prevails in federal-academic ties: Q&A with AAU's Rosenzweig," *Science & Government Report* (February 15, 1992):4–5.

7. This position is most succinctly stated by D. S. Greenberg, "Universities, science, government," *Washington Post* (April 16, 1991):A19.

8. Also see W. Lepkowski, "University research: heyday of science seen as over," *Chemical & Engineering News* (December 9, 1991):4.

9. For thoughtful analyses by observers within the academy, see McCloskey (1991); Menand (1991); D. T. Layzell, "Faculty workloads and productivity," *Chronicle of Higher Education* (April 1, 1992):B3–B4; and D. T. Layzell, "Tight budgets demand studies of faculty productivity," *Chronicle of Higher Education*

(February 19, 1992):B2–B3. On redefining faculty roles in engineering, see Meade (1992).

10. These issues have been analyzed in OTA (1988, chaps. 3–4) and OTA (1989).

11. In the words of OTA (1991:198–199), if an institution is "emulating the research university—beware!"

12. As Skolnikoff (this volume) observes, some politicians argue that if foreign countries are granted equal access to American university research, the United States will lose economically. Neither of the two policy implications of this view, blocking foreign access or declining to fund university research for its spinoffs, is acceptable. Furthermore, the end of the Cold War and declining defense budgets could have an enormous negative impact on the research universities (Sapolsky, this volume).

13. Also see Acton (1991).

14. Also see Browning (1992a) and D. S. Greenberg, "In defense of pork barrel science," *Washington Post* (March 31, 1992):A17.

15. The problem is characterized by competing hypotheses and no decisive evidence. Proponents of megaprojects argue that they have a multiplier effect, i.e., a rising tide lifts all boats; opponents insist that such funding creates a zero-sum game, i.e., the commitment to capital-intensive, multiyear megaprojects leaves less money for small-team, little science projects. For example, see A. H. Teich, "Discussions of setting science priorities are filled with misunderstandings," *Chronicle of Higher Education* (January 22, 1992) and the reply by D. Chubin, "Scientists must enter federal budget fray," *Chronicle of Higher Education* (February 26, 1992):B3–B4.

16. For insight into their contemporary counterparts, see Broad (1985).

17. Also see C. Cordes, "Dominance of science policy by physicists seen as waning with end of cold war and rise of biomedical research," *Chronicle of Higher Education* (March 4, 1992):A1, A32.

18. This is, of course, the still-unfolding tale of the OTA (1991) report, and how the research community has greeted its appearance and responded to its messages. Leon Lederman and the physics community have been most outspoken about how "policy analysts inside Washington" are out of touch with the pain and fiscal travails of researchers throughout the nation.

19. For a rationale for this change in nomenclature, see NSF (1986). Also see W. Lepkowski, "NSF gears up for stricter oversight of academic research support," *Chemical & Engineering News* (January 6, 1992):9–14.

20. For related thoughts of a university chancellor and a provost, and a university provost, respectively, see Atkinson and Tuzin (1992) and Wrighton (1992).

21. For perspectives from OSTP, the agencies, and Congress, see Marshall (1992), Palca (1992), and "Notebook: Thanks, but the answer's no," *The Scientist* (March 2, 1992):4.

22. Universities are urged to acknowledge this, and the federal government is

being counseled to adopt a bolder stance toward developing civilian technology and "precommercial R&D," both through agency programs and alliances with industry. For example, see NAS (1992), Marshall (1992a), Branscomb (1993), and M. Schrage, "Single-company deals are no way for universities to promote research," *Washington Post* (March 13, 1992):D3.

23. As Stanford is discovering in the wake of its overhead problems, the scaling back of its indirect cost rate from 74 to 55 percent has profound effects on its ability to maintain programs and services. See Barinaga (1992a,b) and Hamilton (1991).

24. Derek Bok has said, "Today, university leaders are largely silent, too heavily burdened with raising funds and administering their huge institutions. There is no one able to communicate a compelling vision of what we are trying to accomplish for our students"; quoted in C. J. Mooney, "Bok: to avoid bashing, colleges must take a leadership role on national problems," *Chronicle of Higher Education* (April 8, 1992):A17–A18. Also see Rhodes (1991).

25. The appointment of John Gibbons, formerly the director of OTA, as the president's science advisor has caused a change in OTA director for the first time in almost a generation.

26. See W. Lepkowski, "Congress again explores ways to improve science policy," *Chemical & Engineering News* (October 9, 1989):17–18.

27. See L. Wilson, "U.S. research universities now confront fateful choices," *The Scientist* (March 16, 1992):11.

28. See J. Mervis, "Radcliffe president lambastes competitiveness in research," *The Scientist* (January 20, 1992):3, 7.

29. See D. S. Greenberg, "Science advice in Washington: A male monopoly," *Science & Government Report* (April 1, 1992):1–3.

Bibliography

Abelson, P. H. 1992. Scientific research in universities. *Science* 256:9.

Acton, J. P. 1991. The budgetary environment for science priorities. Paper presented at the Consortium of Scientific Society Presidents, Washington, DC, December.

Atkinson, R. C., and Tuzin, D. 1992. Equilibrium in the research university. Unpublished paper. University of California, San Diego.

Ausubel, J. H. 1992, January. Intellectual migrations and global universities. Unpublished paper. Rockefeller University, New York.

Barinaga, M. 1991. The foundations of research. *Science* 253:1200–1202.

Barinaga, M. 1992a. Stanford and MIT in the dock? *Science* 255:398.

Barinaga, M. 1992b. Stanford faculty tackles overhead. *Science* 255:916–917.

Bloch, E. 1991. Optimists, skeptics, and realists: Other views of the research "crisis." In M. O. Meredith, S. D. Nelson, and A. H. Teich, eds., *Science and Technology Policy Yearbook, 1991*, 147–153. Washington, DC: AAAS Press.

Boyer, E. L. 1990. *Scholarship Reconsidered: Priorities of the Professorate.* New York: Carnegie Foundation for the Advancement of Teaching.

Branscomb, L. M., ed. 1993. *Empowering Technology: Implementing a U.S. Strategy.* Cambridge: MIT Press.

Broad, W. J. 1985. *Star Warriors: A Penetrating Look into the Lives of the Young Scientists Behind Our Space Age Weaponry.* New York: Simon & Schuster.

Brown, G. E. 1992. Scientists must help set priorities. *Space News* (March 2/8):19.

Browning, G. 1991. Testing the hypothesis. *National Journal* (May 4):1078.

Browning, G. 1992a. Colleges at the trough. *National Journal* (March 7):565–569.

Browning, G. 1992b. Forbidden fruit. *National Journal* (October 12):2477–2481.

Bush, V. 1990 [1945]. *Science: The Endless Frontier.* Washington, DC: National Science Foundation.

Carey, W. D. 1988. Scientists and sandboxes: Regions of the mind. *American Scientist* 76(March–April):143–145.

Carnegie Commission on Science, Technology, and Government. 1992a. *Science and Technology in U.S. International Affairs.* New York: Carnegie Commission.

Carnegie Commission on Science, Technology, and Government. 1992b. *Science, Technology, and Congress: Analysis and Advice from the Congressional Support Agencies.* New York: Carnegie Commission.

Chubin, D. E. 1985. Open science and closed science. *Science, Technology, and Human Values* 10(Spring):73–81.

Chubin, D. E. 1987. Designing research program evaluations: A science studies approach. *Science and Public Policy* 14(April):82–90.

Chubin, D. E. 1992a. DEC, OTA, and STS: In praise of marginality. In S. Jasanoff, ed., *The Outlook for STS: Report on an STS Symposium & Workshop*, 45–54. Ithaca: Cornell University Department of Science & Technology Studies.

Chubin, D. E. 1992b. The answer is more questions. Paper presented to the Commission on Professionals in Science & Technology, Preparing for the 21st Century: Human Resources in Science & Technology, Washington, DC, March.

Chubin, D. E., Davies, R., and Heinz, L. C. 1987. Science and society. *Issues in Science and Technology* 4(Fall):106–107.

Cozzens, S. E., Healey, P., Rip, A., and Ziman, J., eds. 1990. *The Research System in Transition.* NATO ASI Series D, vol. 57. Boston: Kluwer Academic Press.

Feller, I. 1992. American state governments as models for national science policy. *Journal of Policy Analysis and Management* 11:288–309.

Geiger, R. 1988. Milking the sacred cow: Research and the quest for useful knowledge in the American university since 1920. *Science, Technology, and Human Values* 13(Summer–Autumn):332–348.

Goldsmith, M. 1986. *The Science Critic.* London: Routledge & Keegan Paul.

Government-University-Industry Research Roundtable (GUIRR). 1992. *Fateful Choices: The Future of the U.S. Academic Research Enterprise.* Washington, DC: National Academy Press.

Hamilton, D. 1991. Indirect costs: Round II. *Science* 254:788–790.

Kantrowitz, A. 1992. Physicists in the "age of diminished expectations." *Physics Today* (March):61, 63.

Koshland, D. E. 1992a. Job application. *Science* 255:513.

Koshland, D. E. 1992b. The race for the "gold" in research. *Science* 255:1189.

Mannheim, K. 1952. *Essays on the Sociology of Knowledge.* London: Routledge & Keegan Paul.

Marshall, E. 1991. Weapons labs: After the Cold War. *Science* 254:1100–1103.

Marshall, E. 1992a. Industrial R&D wins political favor. *Science* 255:1500–1502.

Marshall, E. 1992b. Space scientists heed call to set priorities. *Science* 255:527–528.

McCloskey, D. N. 1991. Invisible colleges and economics. *Change* (November–December):10–11, 54.

Meade, J. 1992. A question of balance. *ASEE Prism* (April):20–23.

Menand, L. 1991. What are universities for? *Harper's Magazine* 283(December):47–56.

National Academy of Sciences (NAS). 1992. *The Government Role in Civilian Technology.* Washington, DC: National Academy Press.

National Science Board (NSB). 1989. *Science & Engineering Indicators, 1989.* 9th ed. NSB-89-1. Washington, DC: U.S. Government Printing Office.

National Science Foundation (NSF). 1970. *Graduate Student Support and Manpower Resources in Graduate Science Education, Fall 1969.* NSF-70-40. Washington, DC: NSF.

National Science Foundation (NSF). 1986. *Final Report.* Advisory Committee on Merit Review. NSF-86-93. Washington, DC: NSF.

Newby, H. 1992. One society, one Wissenschaft: A 21st century vision. *Science and Public Policy* 19 (February):7–14.

Office of Technology Assessment (OTA). 1988. *Educating Scientists and Engineers: Grade School to Grad School.* Washington, DC: U.S. Government Printing Office.

Office of Technology Assessment (OTA). 1989. *Higher Education for Science and Engineering.* OTA-BP-SET-52. Washington, DC: U.S. Government Printing Office.

Office of Technology Assessment (OTA). 1991. *Federally Funded Research: Decisions for a Decade.* OTA-SET-490. Washington, DC: U.S. Government Printing Office.

Palca, J. 1992. NIH's vision runs into political reality. *Science* 255:529–530.

Peters, L. S., and Etzkowitz, H. 1990. University-industry connections and academic values. *Technology in Society* 12:427–440.

Price, D. de S. 1961. *Science Since Babylon.* New Haven: Yale University Press.

Rhodes, F.H.T. 1991. Shaping the future: Science and technology, 2030. *Physics Today* (May):42–49.

Sanders, R., and Brown, F. R. 1966. *Science and Technology: Vital National Assets.* Washington, DC: Industrial College of the Armed Forces.

Savage, J. D. 1991. Saints and cardinals in appropriations committees and the fight against distributive politics. *Legislative Studies Quarterly* 16(August):329–346.

Schmitt, R. W. 1992. The sources of discontent in academic research. Paper presented at the annual meeting of the American Association for the Advancement of Science, Chicago, February.

Schott, T. 1991. The world scientific community: Globality and globalization. *Minerva* 29:440–462.

Shapley, D., and Roy, R. 1985. *Lost at the Frontier.* Philadelphia: ISI Press.

Smith, B.L.R. 1986. *American Science Policy Since World War II.* Washington, DC: The Brookings Institution.

Steelman, J. R. 1947. *Science and Public Policy: A Program for the Nation.* Vol. 1. The President's Scientific Research Board. Washington, DC: U.S. Government Printing Office.

Sweet, W. 1992. DOE advisory committee issues draft report on national labs. *Physics Today* (March):51–53.

U.S. Congress. 1986. *A History of Science Policy in the United States, 1940–1985.* Science Policy Study Background Report No. 1. Task Force on Science Policy, House of Representatives, Committee on Science and Technology, 99th Cong., 2nd sess. Washington, DC: U.S. Government Printing Office.

Weinberg, A. 1966. *Reflections on Big Science.* Cambridge: MIT Press.

White, R. M. 1991. Too many researchers, too few dollars. *Issues in Science and Technology* 7(Spring):35–37.

Wrighton, M. 1992. Testimony presented to the Subcommittee on Science, Committee on Science, Space, and Technology, House of Representatives, Washington, DC, February.

Ziman, J. 1991. A neural net model of innovation. *Science and Public Policy* 18 (February):65–75.

7

Views from the Benches: Funding Biomedical Research and the Physical Sciences

Phillip A. Sharp and Daniel Kleppner

In the United States, "science policy" has been a serious misnomer for what the government does with respect to science. There has been no statement of national goals or objectives; there has been no unified budget for research and development spending. Although there have been many attempts to consolidate the federal government's various roles in science and technology, most recently in the creation of the National Science and Technology Council (NSTC), the institutions, the politics, and the policy for science remain pluralistic. There are, in fact, many sciences, many policies, and little science policy.

Phillip A. Sharp, a biologist, and Daniel Kleppner, a physicist, portray the different experiences of the biomedical sciences and the physical sciences as recipients of federal support. Sharp acknowledges the generosity of steadily increasing support for biomedical science, both historically since World War II, as well as during the past decade. Biomedical research has been buoyed by the interaction of government spending, private investment, and intellectual property. As life scientists have created tools and advanced their research agendas to include more direct study of humans, the relatively straightforward connection between biomedical research and biomedical application has attracted a great deal of public support. Interest groups now lobby Congress for even more funding.

The experience of the physical sciences has been quite different. Kleppner laments that "small science" at universities seems to have been forgotten in the wake of "megaprojects" and political demands for relevance. Even as megaprojects such as the Superconducting Supercollider (SSC) or the Strategic Defense Initiative (SDI) are scrapped or scaled back, the fate of small science does not improve because these large projects have become identified with all of the physical sciences.

Despite their divergent diagnoses, Sharp and Kleppner both suggest that the United States needs more coherent science policy, whether to preserve its international leadership in biomedical research and pharmaceuticals or to prevent its slippage to second-rate status in physics, chemistry, mathematics and materials science. [Eds.]

I. The Biomedical Sciences in Context
Phillip A. Sharp

Federal support for research in biological sciences remains generous and will probably continue to grow in the decades ahead. This strong prospect reflects a highly favorable view of biomedical research by the general public, which desires the benefits of breakthroughs in medical treatment and public health. It also reflects the increasing excitement in biological sciences generated by a shift of focus to the human organism as an object of inquiry. Nothing captures the imagination more fully than an insight into how human beings function or a prediction of how their future may be improved. The popular press has recognized this thirst and has dramatically increased its coverage of biomedical research. Newspapers bristle with articles on the most recent discovery of a disease-related gene, new thoughts on an important biological process, or the introduction of novel therapeutics. With this increased awareness of the potential of biomedical science, a number of advocacy groups have been formed to lobby Congress to increase spending in biomedical research and development (R&D). These advocacy groups speak for the scientific community and research universities and have had a large impact on the growth of funding for biomedical science during the 1980s.

The Commercial Context of Biomedical Research

The current setting for the biological sciences was shaped by events in the early and mid-1970s, when researchers developed recombinant DNA methodologies. These technologies permitted the intentional manipulation of genetic material and thus offered the potential for the genetic engineering of useful, new organisms as well as the accidental creation or release of dangerous ones. The public debate surrounding the development and use of genetic engineering was initiated by a moratorium on such experiments proposed by a small group of scientists. Following the proposed moratorium, an unprecedented meeting of scientists at the Asilomar conference center in California in 1975 discussed conditions or guidelines for commencement of experiments. This highly publicized event led to more formal activity by the federal government. Congress held hearings and debated a number of bills proposed to control recombinant DNA research. Eventually, the National Institutes of Health (NIH) issued a set of guidelines to regulate such experiments. This affair brought to the attention of the public that a new, powerful, and even threatening biology was on the horizon.

Biologists described the new techniques as holding the key to profound medical problems such as cancer and genetic diseases, and as even having the potential to provide new materials and processes for the betterment of life beyond the medical realm. This optimistic assessment had two important consequences: first, it created a climate for tremendous investments by the private sector in commercial opportunities in biotechnology; and second, it fostered public support for increased investment by the federal government as well.

The promise of commercial biotechnology led Wall Street to fund the development of many new start-up companies during the 1980s. These firms were, and primarily remain, highly focused on using the tools of genetic engineering to research and develop medical products. The novel aspect of the meteoric rise of these firms was the availability of private capital even though they were not expected to become profitable for at least five to ten years. Wall Street was able to sell stock in these new biotechnology firms—some $20 billion worth during the 1980s—to both large investment houses and private investors. To date, several of the initial firms have become profitable, some have combined with larger organizations, and some have disappeared. These biotechnology firms have provided employment for many students and have extended and applied the basic science conducted elsewhere. Ideas as well as people have flowed between universities and these organizations to create wealth in the community.

Nevertheless, the commercial success of biotechnology companies, as with the computer and software companies that had been generated by university research in the previous generation, have created some of the current problems for university research. Local and national political leaders have increasingly come to recognize the university as a source of local wealth, and these politicians understandably want to encourage university-based research as a means of regional development in their state or district. This desire to harness the commercial potential of universities has stimulated the earmarking of research funds to local institutions and has reduced the influence of merit review of grants (see Chubin, this volume). Universities that depend upon excellence to attract research support are threatened by this trend.

As universities become more identified with commercial wealth, they also lose their uniqueness in society. They are no longer viewed as the ivory towers of intellectual pursuits and truthful thoughts, but rather as enterprises driven by arrogant individuals out to capture as much money and influence as possible. The media find justification in this new vision of universities and depicts them as institutions of vested interests and privileges. In truth, universities are engines of wealth and are privileged. However, they are also the source of education for the next generation, and this unique aspect of university service is often underemphasized.

The Funding Context for Biomedical Research

The total support for biological and biomedical R&D has grown in real dollars through most of the post–World War II period, but during the 1980s the rate of growth accelerated in the context of declining spending in other parts of the domestic budget. For example, total national expenditures (public and private) for biomedical R&D rose from $9 billion in 1981 to about $24 billion in 1991, in constant dollars using the high biomedical deflator, representing an increase of between $16 and $18 billion using the standard deflator (NIH 1991). During the same period, support from NIH for research grew from $4 billion to $9 billion in current dollars (NIH 1991). This amount compares favorably to the support from the National Science Foundation (NSF) for all areas of science. Specifically, the NSF budget stood at $2.5 billion in 1991, up from $1 billion in 1981 (in current dollars, NSF 1992). Despite this decade of growth at NSF, the total budget is much too small to support the breadth of science under its mandate appropriately.

R&D spending at NIH funds a combination of basic and clinical research. The emphasis has always been—and still remains—on the former, which is carried out in universities, medical schools, hospitals, independent research institutions, and the intramural laboratories of NIH. Hospitals and independent research institutions expanded rapidly during the 1980s and have attracted increasing amounts of support. Typically, scientists in these organizations are not extensively involved in education at either the undergraduate or graduate level and have a majority of their time available for research. In many cases, their salaries are paid almost exclusively from research grants. In hospital settings, the salaries of scientists engaged in research are supported from revenues generated from clinical practice. The expansion of the number of scientists seeking research funds has made the NIH system more competitive for funding of an individual scientist's grant application. Thus, though the NIH budget has grown in real dollars, there are more scientists applying for this money and the success rate per grant or scientist has either remained the same or decreased (see also Vest, this volume). This situation increases the importance of excellence of the faculty in attracting research support and the necessity of assisting younger faculty to become established in the system.

The investment by the private sector in biomedical R&D is of the same scale as that of the NIH budget. However, some 80 percent of this money is committed to product development and clinical trials rather than to the type of basic research performed in universities. Only a small fraction of the remainder is spent by companies for support of basic research outside their own laboratories. The basic research performed within firms is frequently used to allow industrial scientists to remain in contact with research done in university laboratories and to induce the flow of ideas and information between the private and public sectors. Thus, the pharmaceutical industry relies significantly on the basic research supported by NIH to complement its own R&D.

One of the few large industries in which the United States remains a leader is pharmaceuticals. The continued success of the U.S. pharmaceutical industry in the international marketplace is one indicator that this public investment in biomedical R&D has been productive. But given the size and wealth of the biomedical research community in the United States, it is not surprising that many international pharmaceutical firms such as Sandoz, Glaxo, Hoffman-La Roche, and BASF have recently decided to expand their research laboratories in this country. The possibility of foreign firms "free riding" on public R&D spending in universities that this trend may indicate has caused consternation in Congress and reactions by some universities (see Skolnikoff, this volume).

The historic growth of the NIH budget over the past four decades can be attributed to the intense political activity of a number of nonprofit groups and interested individuals, often referred to as the "disease lobby" because they commonly are committed to the treatment or eradication of a particular disease, such as cancer or cystic fibrosis. Mary Lasker and Benno Schmidt, working together with friendly forces in Congress, helped build NIH and initiated the War on Cancer by lobbying for the passage of the National Cancer Act of 1971. This act provided funds for the expansion of biology departments and the development of laboratories dedicated to the War on Cancer, such as the Center for Cancer Research at the Massachusetts Institute of Technology (MIT). Other interest groups have also worked with Congress and the administration to increase funds for their particular disease. Most recently, this pattern has been repeated by the political action groups supporting research on acquired immune deficiency syndrome (AIDS). Congress increased the NIH budget by more than a billion dollars with a commitment to treat and control the AIDS epidemic.

These interest groups are often effective advocates for support of research because they are not identified with university scientists, and thus their calls for increased R&D spending do not appear to be self-serving. Their message to Congress and the administration has advocated an increase in research funds for scientists and physicians who can help the public or a large segment of it. This public-regarding justification for biomedical science can be contrasted to the recent plea for an increase in funding by physicists—

led by Leon Lederman—which was met with public and political disfavor because it appeared self-serving.

A recent example of a national organization whose objective is to increase federal support for biomedical research is Research!America, founded by Jack Whitehead, the benefactor of the Whitehead Institute for Biomedical Research at MIT. Research!America's public campaign has taken advantage of popular concern over health care, emphasizing that the total level of R&D expenditures account for only 3.8 percent of the total cost of health care and arguing that more research should reduce the cost and increase the effectiveness of health care.

Research!America has also worked to document public support for increased funding for biomedical research, as demonstrated by its recent survey of registered voters in Maryland (Haynes, Smith, and Winogradzki 1992). Over 60 percent of the sample responded that too little money is being spent on medical research. Only education, health care, and the environment were ranked higher for an increase in funding. Similarly, over 77 percent of those polled do not feel that governmental programs go far enough in solving problems in medical and health care fields. The combination of interest in the field and prospects for future developments is reflected in the answer to the ranking of possible careers for young people: over 80 percent of those polled would suggest medical researcher as a career, a favorable rating exceeding even that of a physician. A vast majority, 77 percent, also believes it is very important for the United States to maintain its role as a world leader in medical research. Somc 46 percent of the sample felt medical research was very important for the economy of Maryland. The depth of this feeling was reflected in the fact that 47 percent favored increasing federal spending for medical research even if it meant increasing taxes. Research!America underwrote the development of the poll to provide evidence to the administration for the public's support for medical research. This kind of favorable public opinion, combined with an obvious human need, are strong inducements for increased support for biomedical research in the future.

The next decades will probably bring a decrease in the country's willingness to spend money on defense and military-related activities. The three major political issues of the future will be health care, the environment, and the economy. There is no doubt that biological sciences are critical for progress on the first two issues and is becoming increasingly important for the third issue. Universities have recognized this trend and have expanded programs related to biological sciences.

At MIT, this recognition has taken the form of an expansion in the Department of Biology with both the Center for Cancer Research and the Whitehead Institute for Biomedical Research. Beyond the Department of Biology, MIT has developed the Whitaker College, with its Health Sciences and Technology program that blends doctoral research and medical education, and the Brain and Cognitive Sciences program. The Departments of Chemistry, Toxicology, Civil Engineering, and Electrical Engineering all have programs in the biological sciences. Furthermore, the faculty of MIT have decided to require a core course in biological science of all undergraduates. These programs will likely grow in importance at MIT in the future. One hopes that the foresight these programs represent will be common among other research universities and in the federal government in anticipation of the increasing role of the biological sciences in society.

Bibliography

Haynes, D., Smith, H., and Winogradzki, A. 1992. *Medical Research and Health Care Concerns: A Survey of Maryland Voters.* Alexandria, VA: Research!America.

National Institutes of Health (NIH). 1991. *NIH Data Book.* P. E. McKinley, ed. Bethesda: NIH.

National Science Foundation (NSF). 1992. *FY1993 Budget Request.* Washington, DC: NSF.

II. Funding the Physical Sciences

Daniel Kleppner

Research in the physical sciences is carried out in relatively small laboratories at universities and in large, off-campus facilities. The contrast in organization, operation, and funding between these modes of research is enormous. The consequent debate between "small science" and "big science" generally emerges as a central theme in any science policy discussion, particularly in the physical sciences. Arguments over the relative priority that these modes of research deserve have at times been rancorous, and they threaten to become more rancorous as requests by scientists continue to outstrip available resources. A few scientists take the extreme position that big science is simply not worth the cost, and that in times of financial stress, one should scrap "megaprojects" and concentrate resources on small groups and small science. Such a view is excessively narrow and overlooks the historical fact that as long as there has been science, there has been big science.

Big Science

With only a slight stretch of the imagination, one can trace the origins of big science to ancient times when science and religion had not yet parted ways. The monoliths of Stonehenge form a calendric instrument as well as a religious monument. The enormous stones were transported hundreds of miles to the site. Whether the project was completed on time and on budget must remain a matter for speculation.

From the calendric carvings emerged the power to predict the passing of the seasons, the phases of the moon, and the occurrence of eclipses. Such power engendered the growth of astrology, and from astrology came astronomy, the first modern science. The progression from Tycho Brahe's observatory to William Herschel's giant telescope, to the Mt. Palomar observatory, to the Hubble space telescope is all of a piece. However, when the specter of big science is raised today, it is not the vision of the solitary astronomer at the controls of a large instrument that comes to mind, but rather numerous scientists, technicians and students working on a fusion machine, a particle accelerator, or perhaps even a space station. To the public, and sometimes to their representatives, big science is science itself. The failure to distinguish between the broad base of fundamental research in our universities and these scientific megaprojects seriously distorts the picture of science and science policy in the United States.

The rationales for scientific megaprojects generally fall into one of three categories. The first is the pure and simple pursuit of new knowledge. The Hubble space telescope and the Human Genome

Project are primarily motivated by this goal, as was the late Superconducting Supercollider (SSC). The second category of megaprojects addresses a specific national need, for example, the fusion program to meet national needs in energy, or the global monitoring program to meet national (and international) environmental needs. The third category encompasses research devoted primarily to achieving political goals, such as the former Strategic Defense Initiative (SDI) and the space station (Press 1988).

The social payoffs from basic research—regardless of scale—are unpredictable and may not be realized for decades. Such a time frame is too long to suit most congressional temperaments, and so megaprojects are often justified by exaggerated claims of a more immediate nature. For example, the following rationale was provided by a Senate committee report for the SSC:

> The Committee recognizes the importance of the SSC to the Nation's technological health. More than just an investment in U.S. scientific leadership, the SSC is the impetus for new directions in industry, education, and economic growth. Over 100 American universities and nearly 900 American scientists are preparing to do research utilizing the SSC. By the end of fiscal year 1992, SSC educational programs will have directly reached more than 30,000 students and educators. More than 6,000 jobs across the country have been created by the SSC, including new opportunities for engineers, technicians, construction, and other workers experiencing reduced job opportunities as a result of cutbacks in the defense industry. (U.S. Congress 1992)

If the issue were how to invest $8 billion to stimulate education and create new jobs, there would be more effective ways to proceed than by constructing a mammoth accelerator in rural Texas. Nevertheless, had the SSC gone forward, there is good reason to believe that over time it would have paid off by generating new knowledge, sustaining interest in science, and possibly even generating new technologies. The exaggerated claims for near-term economic payoff, however, do not appear to have played an important role in its demise. Congress withdrew support for the SSC primarily because of the increasing costs—over $11 billion at the time of the vote—and the lack of international participation. In addition, the vote expressed a growing impatience with basic research among members of Congress. That same impatience has been similarly expressed in recent efforts to micromanage the National Science Foundation (NSF) through the appropriations process, especially calls for NSF to focus its efforts on "entrepreneurial" activities and the transfer of technology from academic laboratories to the private sector.

The largest megaproject currently on the drawing board is the space station, which has already consumed roughly $10 billion and is currently spending about $2 billion per year. Its goals have shifted considerably from its original objective of space science. Most scientific experiments in space are much cheaper and quicker on unmanned vehicles, a point made forcefully by the Augustine Report in 1990 (Advisory Committee 1990). The space station, however, has been reconfigured to eliminate many of the planned experiments. Its goal was similarly redefined to study whether humans can survive for long periods in space, preparatory to a manned mission to Mars, and to collaborate with Russia in the hopes of keeping some of the space scientists of the former Soviet Union gainfully employed. Although its actual scientific program is minimal, the National Aeronautics and Space Administration (NASA) continues to portray the space station as a science-driven project. But just as it is difficult to think of more ineffective ways to educate and employ people than the SSC, it is similarly difficult to think of a less cost-effective way to pursue space science than on an inhabited space station. Nevertheless, to the public, the space station serves as a prominent symbol of the nation's commitment to science.

Because of the high visibility of scientific megaprojects, Congress and the public view the nation's commitment to science as strong and its scientific prospects as bright. Such a vision is deceiving, however, for the future of science rests primarily in the nation's universities, and the prospects for university-based research, at least in the physical sciences, are by no means bright.

Physical Sciences at NSF and the Universities

In the programs of the NSF, the physical sciences include mathematics, chemistry, physics, astronomy, and materials research.

They are grouped together in NSF's Mathematics and Physical Science Directorate (MPS). In these disciplines, NSF provides the largest segment of all federal support; for example, NSF provides more than half of the federal support for basic research in physics. Thus, a portrait of MPS activities is to a large extent a portrait of basic research in the physical sciences at the nation's universities.

In January 1987, President Ronald Reagan made two significant proposals for science policy. He announced the commitment of his administration to building the SSC, and in his Message to Congress declared that the NSF budget would be doubled in the following five years (Reagan 1989). Today, the nation is now debating how much money to spend to decommission the large hole in the ground where the SSC was to have been. Although not achieving President Reagan's goal, the NSF budget performed significantly better: it grew from $1.6 billion in fiscal year 1987 to $2.6 billion in 1992, an increase of 63 percent that is remarkable considering the problem of the deficit.

Under closer scrutiny, however, the picture of the NSF budget represents not an enhancement, but in many ways an abandonment of basic research in the physical sciences. The budget growth at NSF was allocated primarily to engineering, technology, and education. The MPS budget remained flat at approximately $620 million, in constant 1992 dollars. Furthermore, these constant dollars are calculated with the standard consumer price index, which does not fully account for the inflation of the costs of research (see Chubin, this volume). As a result, the physical sciences are significantly worse off than they were in the mid-1980s.

The picture of the physical sciences funded by NSF is bleak. Excellent proposals are routinely turned down. Grants are often much smaller than the actual costs of research. For example, the median size of a grant in the physics program is $81,000, but the estimated cost of running a small experimental group in physics at the Massachusetts Institute of Technology, for example, is $225,000. Researchers must therefore rely on several grants, resulting in excessive time devoted to fund-raising activities. Young researchers have great difficulty getting started, while more established scientists whose work receives outstanding reviews frequently have their grants cut. Facilities and equipment are often obsolete.

In short, university-based research in the physical sciences has been slipping into disarray.

Facing the Future: Science without Science Policy?

Federal expenditures on research and development (R&D) are currently about $75 billion annually. Roughly $1.2 billion of this total is spent on university-based research in the physical sciences. A small redirection of spending priorities within the overall R&D budget could easily reverse the downward drift of the physical sciences in the universities. However, it seems unlikely that such a redirection will happen, because short-term political considerations outweigh the nation's long-term interests in investing in university-based physical science.

The proper balance of spending between big science and little science has been frequently debated. However, such debates generally amount to wasted breath because the nation lacks any coherent forum for addressing the issue. Not only this issue but the great majority, perhaps all, of our crucial policy issues are being dealt with on an essentially ad hoc basis. The list includes the nation's major long-range objectives in science, the relative importance of science for its own sake and science as a force for economic development, scientific elitism versus the traditions of democracy, the desirable size of our rapidly growing scientific community, the restructuring of our institutions in response to the changing national and international situation, and more. The list is long and many of the issues have been addressed by other authors in this volume.

Underlying this nation's past scientific success is a tradition of scientific pluralism. Scientists have been free to turn to many agencies, each with its own research agenda. For several decades the system worked well, allowing researchers to move rapidly and the nation to avoid the pitfalls of rigid and shortsighted policies. However, for over a decade the system has been working poorly. Even as the nation scales back plans for megaprojects, the individual investigator is becoming an endangered species. The time has come when the nation must either rethink how to maintain scientific excellence, or prepare to add science to the long list of activities in which the United States was preeminent but is now in the second rank.

Bibliography

Advisory Committee on the Future of the U.S. Space Program. 1990. *Report to the Vice President.* Washington, DC: The White House.

Press, F. 1988. The dilemma of the golden age. Address to members, 125th annual meeting of the National Academy of Sciences, Washington, DC.

Reagan, R. 1989. Message to the Congress on "A Quest for Excellence." *Public Papers of the Presidents of the U.S., 1987,* 61–79. Washington, DC: U.S. Government Printing Office.

U.S. Congress. 1992. *Energy and Water Development Appropriations Bill, 1993.* Senate Appropriations Committee. Senate Report 102-344, 102nd Cong., 2nd. sess., July 27. Washington, DC: U.S. Government Printing Office.

8

Financing Science after the Cold War

Harvey M. Sapolsky

Calculated not only to defeat the Japanese but also to limit Soviet participation in the Pacific peace, the atomic bombing of Hiroshima and Nagasaki was one of the first acts of the Cold War as well as one of the last of World War II. The atomic bomb thus became the pivotal symbol of the transition from World War II to the postwar era. As a symbol, it embodied the ambivalences of the era: the awesome, instrumental power of science in service of the state; and the similarly awesome power of knowledge to defend freedom against tyranny. A decade or so into the postwar era, the atomic bomb was wedded to that other technological symbol of the Cold War, the missile, which similarly embodied stark ambivalences: conquering intercontinental distances to deter the implacable foe; and conquering interplanetary distances to open a new frontier. So powerful were the bomb and the missile, as instruments and as symbols, that their politics—of races, gaps, and control—became the central politics of the Cold War. And as Harvey M. Sapolsky shows in this chapter, that politics was reflected to the relationship between government and the research scientists who in their pursuit of knowledge had wrought these terrible, wonderful, necessary things.

Now the Cold War has evaporated, leaving behind institutions, people, and funding patterns that had been the solutions to old problems. Sapolsky suggests that the impact of the Cold War's demise on research science will be uneven. It will not be devastating to universities, which weaned themselves from most defense spending a generation ago, but it may decimate specific fields that had come to depend on military funding. The major impact Sapolsky anticipates is "the democratization of science"—that is, the increased geographic dispersion of research funding under the increased influence of domestic political considerations. In his formulation, the Cold War clause of the government-science contract has come undone, and this has made the other clauses of the contract that much more fragile. [Eds.]

Introduction

What we all said we wanted to happen, but had given almost no thought to, has indeed happened. The Cold War has ended, and fortunately so with liberty in triumph and without much blood having been shed. As we watched, the tyranny in the East crumbled morally and economically, freeing millions of people. In fact, of course, for most U.S. citizens the Cold War in its fifth decade was not a very demanding or dangerous conflict. We had become used to the modest mobilization it required of society and so had most of our institutions. The war gave order to the world and money to many of our favorite causes, if not always with honest justification. The personal risk for most of us, no matter what was claimed by either the antinuclear movement or defense planners, was almost nonexistent. On the contrary, the war was a comforting part of life like the Social Security system, not quite what we might want it to be, but something we found reassuring to know was there nevertheless.

It is not surprising then that the rapid end to the Cold War would be greeted with generalized anxiety. We suddenly discover that there are important adjustments to be made, but realize that there is no big, well-discussed plan for these adjustments. Obviously, they will be much more painful to achieve in the East than the West, but even here there are difficult changes to be made. Some regions of the country, some industries, and some organizations are very dependent upon the military. Although we do not yet know precisely at what level defense spending will settle, we do know that there will be substantial reductions, and that the cuts cannot be avoided for very long.

Overall, the ending of the Cold War will not have a great effect on university research. This is a sector that was weaned relatively early from defense, mostly in the 1960s and 1970s. No more than 10 percent of U.S. university research is currently financed by the Department of Defense. And even after the reductions occur some percentage, perhaps most, of this spending will remain intact if only as insurance against bad people, bad times, and the wars of the future.

But some segments of academic science will feel the effects of a declining national concern about defense. Fields such as aeronautics and oceanography, where the military involvement was intense, are likely victims of changing priorities. Less obvious is the impact the decline will have on the social structure of U.S. science. The current institutional elite in science came to prominence with the initiation of the federal government's patronage of university research, a patronage that was largely founded upon and then long sustained by the Cold War's persuasive national security rationale for the support of science. As the rationale fades in its ability to elicit federal funds, so too do the prospects of at least some major institutions. Alternative rationales exist for science. Some will no doubt attract considerable allocations. But all favor a more equitable distribution of research rcsources than that supported by a national security justification. The democratization of science seems likely to lie ahead.

This chapter examines the consequences of a changing rationale for federal support of science. It begins with an analysis of federal budgetary trends, focusing primarily on defense research and development allocations. Then it explores the distributional effects of the security rationale and other plausible rationales for the support of science. The chapter ends with a description of the politics that will likely rule science in the post–Cold War era.

Technology Mobilized

Research and development activities in the United States have been greatly influenced by national security requirements since the 1940s. The contribution of scientists and engineers to victory in World War II, although easily exaggerated,[1] was the inducement for continuing military attention. The atomic bomb was the most spectacular technological advance made during the war, but there were many other significant developments achieved in aircraft design and propulsion, radar, submarine detection, weapon fusing, and medical care as well (Kevles 1988: 120). The nation's leaders quite readily accepted the argument that technology could be the competitive advantage in any future military conflict. This view was reinforced in the postwar contest for international domination because the Soviet Union was thought willing to use its obvious numerical advantage but to lag badly in technology.

Hundreds of billions of dollars were invested in the effort to maintain the country's technological edge in weaponry. The Cold War saw three peaks in defense spending, recognized easily as the Korean, Vietnam, and the Reagan buildups. Each peak reached approximately the $340 billion level, measured in constant 1992 dollars. The troughs were at about $250 billion, with allocations somewhat higher in the 1950s and 1960s and somewhat lower in the 1970s. Research and development expenditures grew as a percent of total defense spending over time, reaching as much as 15 percent in recent years, but rarely falling below 10 percent at any point in the Cold War. Defense research and development expenditures, measured in constant dollars, were thus nearly always in the $20–40 billion range annually for the past forty years (Lewis 1989, 1991).

To give perspective on these numbers, it is important to note that total research and development expenditures annually in the United States are currently about $150 billion, half of which are provided through federal appropriations. The Department of Defense (DOD) accounts for nearly 60 percent of the federal share, or about 30 percent of total research and development funds in the United States. This mammoth commitment to improving defense technology persists several budget cycles after the Berlin Wall was breached and then dismantled. In the early years of the Cold War the defense share was even greater when nondefense federal research agencies did not receive significant appropriations.

Using DOD expenditures as the measure of defense-related research and development allocations, in fact, underestimates the role of national security rationales in national research and development activities. The five leading federal agencies supporting research are DOD, the National Aeronautics and Space Administration (NASA), the Department of Health and Human Services (HHS), the Department of Energy (DOE), and the National Science Foundation (NSF). Together they account for over 90 percent of total federal research and development expenditures.

DOD obviously has relied on security rationales to gain the bulk of its research and development allocations, but so have three of the other four agencies. NASA was formed in response to the Sputnik challenge, to carry the flag into space and to rebuild worldwide confidence in U.S. technology. DOE is the partial inheritor of the Atomic Energy Commission, and most particularly of the laboratories that develop and test nuclear weapons for the military. DOE's other elements also use a security rationale when they highlight in budget discussions the potential strategic vulnerabilities caused by a dependence on foreign energy sources. Although NSF was created in 1950 with the mission to support academic research, it did not receive substantial appropriations until the late 1950s, when its advocates, some would say with the cynical encouragement of the Eisenhower administration, linked the nation's embarrassment in missile and satellite technology to the national investment in science. Only the research at HHS did not gain appreciably from the defense largess during the Cold War.

Although in decline since 1986, the defense budget has not yet fallen to the levels of the late 1970s, when there was still a Warsaw Pact facing NATO and great tension between the United States and the Soviet Union. It has been an unusually slow drawdown from a cyclical peak in the DOD budget. In large part this is due to the Bush administration's desire to maintain all its military options as the hard-to-predict events unfolded in the East. The Gulf War also delayed reductions, requiring as it did the deployment of a major expeditionary force. Buttressing all the caution and the delays was the budget agreement of 1990, which many now describe as a total debacle for President Bush, but which at the time was considered one of his few domestic political triumphs (Gleckman 1991; Jones 1992).

The budget agreement is best known as the moment when President Bush abandoned his adamant "no new taxes" pledge. But it was also the point at which he removed all possible budgetary and program initiative from the congressional Democrats, preventing them, at least until 1993, from reaping any political leverage from the end of the Cold War. Under the terms of the agreement, the budget shares of defense, domestic nondefense, and foreign assistance—the areas of discretionary spending in the federal budget—

were fixed through fiscal year 1993, barring a declaration of emergency (Hager 1992). Attempts to increase spending on a program within one of these categories required a corresponding reduction in spending on another program within the same category. Increases in entitlement programs are allowed, but only with an increase in taxes. There could be no transfer of spending from defense to domestic spending (or vice versa) until October 1993, well after the 1992 elections. Although the Democrats ultimately won the presidency and retained control of both houses of Congress, they gave up in the agreement the ability to trade defense savings for nondefense spending for three years. And as the agreement did little to reduce the deficit, the Clinton administration has had to follow fiscal priorities that hardly can be distinguished from those of either Bush or Reagan.

Soon, however, there will have to be a substantial adjustment in defense spending. Caspar Weinberger, who served as Secretary of Defense through most of the Reagan administration, required the military to plan as if there would be a constant expansion of the defense budget. As a result, essentially all fiscal discipline was abandoned. Many more weapons were programmed for procurement than can now reasonably be expected to be accommodated in the Clinton administration's proposed budgets. If the defense budget does turn down drastically, as seems quite likely in a world in which even former chairman of the Joint Chiefs of Staff General Colin Powell was able to see no "demons," then the number of canceled military development projects will be substantial.[2] The enhancement of national security was the rationale that convinced the nation to support with public funds a very large research and development enterprise in the years following World War II. At various times during this period over 1 percent of the gross domestic product (GDP) was devoted to the cause. But until another military threat as visible and potentially dangerous as the Soviet Union reappears, the defense budget itself may not command very much more of GDP. Historically the peacetime norm for defense (the maintenance of the military) is between 1 and 2 percent of GDP (Warden 1989). Although President Clinton and senior defense officials have promised to keep up the nation's technological guard, it is obvious that the defense contractors, the arsenals and depots, the reserves, and the remaining active duty forces will be competing intensively for favored access to a diminished pool of resources (Kosiak and Taibi 1992). Science and engineering activities in the United States surely need to find other rationales for their support if they are to prosper in the foreseeable future.

The Benefits for Science

University research benefited greatly from the Cold War. Although universities were allocated but a minor portion of defense R&D funds, the amounts universities received were never trivial relative to those available to them for the support of research from other sources. In the initial years of the Cold War especially, the military financed much of the university-based basic research and a considerable amount of graduate training in the sciences in the United States. But even more significant, the military helped define a government-university relationship that extended beyond defense-related work and that persisted largely intact and unchallenged until quite recently.

Military R&D spending is concentrated on development activities and not on research activities, either applied or basic. In fiscal year 1992, for example, DOD agencies spent nearly $43 billion on R&D activities, all but about $5 billion of which was for weapon development and weapon testing, for things that find targets and destroy them. What DOD seeks are new weapon systems, not the advancement of fundamental knowledge. As a result it spends its funds primarily to support work being conducted in industry and, to a lesser extent, in government laboratories and test ranges rather than in universities (NSB 1991: 95–100). Fundamental knowledge is not considered by the military to be irrelevant to weapon development, but only insufficiently early and predictable in application to merit substantial military investment.

DOD does support some basic and applied research conducted at universities, accounting currently for about 10 percent of all such support (U.S. Congress 1986; AAAS 1992). This work is concentrated in particular fields. The military has long supported research in flight dynamics and propulsion, electronics and computers. The

Navy takes a proprietary interest in oceanography and ocean engineering. And various defense-related agencies remain active in financing nuclear and other branches of physics.

In the early post–World War II years, the involvement of the military in university-based research was proportionally greater although not entirely intentional. The military sought to continue work begun during the war and, in some instances, this included supporting research groups that took up permanent residence in or returned to universities at war's end. Later, in the push to acquire ballistic missiles and otherwise enhance military capabilities, additional defense laboratories were established, many affiliated with universities in large part to entice scientists and engineers to join them. But the key impetus to defense support of university research was accidental, arising out of machinations within the Navy immediately after the war to gain control over the project to develop nuclear propulsion for submarines (Sapolsky 1990). Vice Admiral Harold G. Bowen, one of the contestants for directorship of the nuclear-powered submarine project, created the Office of Naval Research (ONR) as his management vehicle, providing for it the bureaucratic cover of an agency with the mission to support advanced research in nuclear physics and other topics of interest to the Navy. The admiral failed in his quest and was forced to retire, leaving behind an organization that was well funded and staffed by scientists who were convinced that the Navy and the nation would benefit by supporting the nation's best science.

The vision that the ONR staff had for the organization in 1946 when it was formally created was much like the one that Vannevar Bush, the leader of the wartime science effort, expressed in 1945 in his self-solicited report to the president on the future support for science titled *Science: The Endless Frontier*. Both the civilian and the military needs of the nation would be well served by large investments in university-based scientific research, Bush had argued. But NSF, the organization he proposed to make the investment, was not established until 1950 and did not dispense significant funds until the end of the decade.[3] In the interim, ONR acted as the federal patron for academic science.

The major research universities should consider it extremely fortunate that ONR and not NSF initially held the patron role.

Senior naval officers did not approve of ONR's role, but they were too preoccupied with first the demobilization following World War II and then the start of the Cold War and the Korean mobilization to take the steps necessary to alter it (Carter 1966).[4] But outside the Navy, ONR was viewed as conducting a military mission, one apparently buttressed by a vital national security rationale. After all, it was the Office of *Naval* Research. ONR thus had the political freedom to negotiate generous terms of support for universities. Easily bypassed were all the traditional constraints on government support such as the barriers to the award of public funds to institutions with religious affiliations and the need for grant recipients to share costs with the government. The GI Bill financed all postsecondary training—trade, professional, as well as undergraduate and graduate—but ONR gave focused assistance to research and graduate training in the nation's leading science and engineering departments.

The arrangements could not have been better from the perspective of scientists and administrators. ONR staff officials were determined to assuage fears that government sponsorship of university research would be restrictive and capricious, always burdened with petty bureaucratic rules and subject to the vagaries of politics (Jewett 1971). Scientists were encouraged to propose their own projects and were not required to submit progress reports—mere publication in the open, refereed literature was considered sufficient. They could fund graduate assistants and their own summer salaries from the awards, which were often multiyear in length and renewable. Administrators were pleased to learn that ONR would follow the special wartime policy of paying full overhead costs, generously defined, thus reimbursing universities for sponsored research at rates that government agencies in the 1930s shunned and private foundations to this day refuse to do (Bush 1970: 38). Literally hundreds of millions of dollars have been provided to research universities for allocation by their administrators via this mechanism.

But the most important advantage ONR's involvement offered university research was its national security rationale, one that provided protection, at least during the Cold War, from the pressure to utilize geographic distributions and similar formulas in research awards. Equity, especially geographic equity, is a powerful tradition in U.S. politics. The nation's federal system often seeks to level differences among the states. Normally program funds have to be dispersed widely, giving all local governments, organizations, or individuals with any imagined hope of qualifying for a federal award a share of the pie. Elite institutions do not do well in this system. Yet ONR was able to concentrate its funds in a relatively few universities, recognizing and reinforcing an establishment with public monies. This capacity to ignore cherished political norms soon extended beyond ONR.

A 1947 debate between Senator Leverett Saltonstall, Republican of Massachusetts, and two Democratic Senators—Richard Russell of Georgia and William Fulbright of Arkansas—over an amendment to the then-pending NSF bill shows the consequences that flowed from having a military agency rather than a civilian agency involved in allocating funds for academic research (Sapolsky 1990: 121–122). The amendment would have required that a geographic distributional formula be used in the allocation of the foundation's grants. Senator Saltonstall argued against the amendment, citing the vital role MIT and Harvard played in various weapon research projects. Senator Fulbright reminded his colleague from Massachusetts that the issue before them involved "pure science research," not applied research, and thus more out-of-the-way places, as he put it, might also be able to contribute.

Senator Russell followed with a prophecy: "I unhesitatingly put in the Record the prediction, here and now, that within six years after the pending bill shall have been enacted into law, in the absence of an amendment of this nature, the two institutions referred to will be receiving more funds than all the educational institutions in at least the 12 states east of the Mississippi River and south of the Potomac."[5] The bill passed without the amendment, only to be vetoed by President Truman on other grounds. Adjusting the dates for the later establishment of NSF, Senator Russell correctly foretold the future. The senator from Georgia, however, also served on the Senate Appropriations Committee, where he routinely approved funds for ONR which initially took the place of NSF and which gave much of its largess, safely protected by the presumption that weapons of some sort were being built, to MIT, Harvard, and a few other universities, none of which was located in the states bounded by the Mississippi and the Potomac.

Breaching the Walls

Although it was long in the making, the end came quickly for the Cold War. The time between the breaching of the Berlin Wall to the disestablishment of the Soviet Union, the empire it protected, was just over two years—26 months to be exact. Already the armies that eyed each other across that wall are fading, their weapons slated either for scrap or storage. It is difficult to believe it is totally over and so quickly.

But we also forget quickly. Many of the things the United States did to enhance its military power during the Cold War seem so strange now. Nuclear-armed bombers stood 15-minute runway alerts, ready to strike Soviet targets. Submarines regularly tested Soviet waters. Eavesdroppers listened to as many Soviet telephone conversations as possible. The country spent lavishly to acquire those bombers, submarines, and satellites, ignoring the customary suspicions of contractors, industrial and otherwise.[6]

Today, the government auditors, some even from ONR, cannot understand why Stanford University officials would believe they could include flowers for their president's home as an overhead charge on research contracts, or why MIT officials would think that limousine service for trustees was an allowable cost.[7] With the war over, there are rules to rewrite and political sensitivities to recalibrate.

The effect of the Cold War's end is seen most clearly in the effort of members of Congress to earmark portions of the federal research budget, but especially that of the DOD, for colleges and universities within their districts. Earmarking is the process of requiring by law that a particular institution (or sometimes a set of institutions) receive a specific award rather than allow an agency to select the most qualified applicant from a pool on the basis of announced criteria, most usually technical abilities or peer judgments. In fiscal year 1988, earmarking amounted to $225 million by one calculation. Two years later it was just $270 million. But in fiscal year 1993 it soared to over $760 million as the defense research budget became fairer game for this form of congressional patronage (Marshall 1991, 1993).[8]

The budgetary instructions are usually quite specific, even to the point of absurdity. For example, DOD was mandated in the fiscal year 1992 budget to provide Marywood College with $10 million for the development of an institute to study stress in military families. The fact that the Scranton, Pennsylvania, college, which is administered by the Sisters of the Immaculate Heart of Mary, had not done much research in this or any other field was to be of no consequence to DOD (Browning 1992). In the same budget, the Navy was told to give the University of Mississippi one million dollars for its "National Center for Physical Acoustics." A few years ago, the Air Force had to provide the funds to construct an engineering laboratory building at the University of Nevada at Las Vegas, which is known more for its winning basketball team than for its work in hydraulics or electronics.

Not all the interests being expressed are geographic. Massachusetts, which has done quite well in the competitive grants arena, also has done well in the earmarking arena, thanks no doubt to some important committee assignments held by members of the Massachusetts congressional delegation (NSF 1989).[9] But in this case the money goes to Boston University, Clark University, the University of Massachusetts, or Northeastern University rather than Harvard or MIT, the usual destination for federal research funds in the state.

Attempts to alter distributions when research budgets are slow growing or stagnant are expected, but they can exist even when such budgets are expanding, as has been the situation recently. Science faculties replicate themselves, generating not only future replacements but also current rivals for resources as students graduate and go off to positions at other institutions. Although the traditional centers for research have expanded, most of the opportunities are at the academic periphery, in the new state universities or revived private ones that have accommodated the growth in undergraduate enrollments that has occurred since the 1960s, and that have acquired ambitions to improve their academic lot.[10] Given its inclination to reproduce, university research is impossible to satisfy. Science, in this sense, is more a bottomless maw than an endless frontier. Neither the academic establishment nor the national treasury is safe from the hunger of the young and the expectant.

The Democratization of Science[11]

The rationales that garner the most support for science now focus not on national security but on health and economic growth. Scientists cultivate, or at least do not discourage, expectations that increased allocations for university research will lead to significant improvements in health status and the economy. The problem for science is that in the long run these rationales protect its independence much less than do national security rationales. Whereas politicians are willing to defer to the judgment of the military on most research matters, they have little hesitation to substitute their own for that of civilian officials in nearly all domestic affairs.

Some civilian science allocations are naturally pleasing to members of Congress. The big budget of the National Institutes of Health (NIH) divides fairly well geographically because most of the money goes to medical schools that are distributed according to population in order to meet their need for "clinical material," training cases for their students. Moreover, first-rate medical schools exist in every region of the country, attracting federal research grants with ease. Boston and San Francisco are major centers for medical research, but so too are Atlanta, Baltimore, Houston, Seattle, Salt Lake City, St. Louis, and two dozen other cities. A multibillion dollar budget assures that no area need feel neglected.

But if geography is not an issue in health research then the content of the research programs can be. At the urging of interest groups, Congress is constantly mandating special attention for particular diseases and disciplines within the NIH budget. NIH prides itself on its control through peer review of specific research awards, but it could not resist demands for a major initiative in cancer that went substantially beyond what most cancer specialists thought appropriate at the time (Rettig 1977; Rushevsky 1986). A program in artificial heart research was begun and recently was renewed on the clear threat of intervention by congressional supporters (Maxwell, Blumenthal, and Sapolsky 1986; Culliton 1988). The allocations for AIDS research now rival those for cancer and heart disease. Women are now complaining about the neglect of their interests in the allocations (Palca 1991). NIH continually searches for a method for determining priorities other than by politics, but never finds it (IOM 1984; Ginzberg and Dutka 1989).[12]

The physical sciences suffer from political pressures as well. The siting of large-scale research facilities such as particle accelerators and electronics centers become the prize in struggles among regions. The Southwest gets what the Northeast and the Midwest wanted. Scientists have learned to encourage the competition by promising economic gains for the winning location. The earmarking of awards, noted previously, is part of this competition. Calls reminiscent of those made by Senators Russell and Fulbright are being heard for limiting the amount that universities in any state could receive in research distributions (Walsh 1987).

Attempts to contain the pressure to spread the scientific wealth are undermined both by the desire of particular universities to improve their comparative standing and by the widespread belief that science enhances economic prosperity. Many universities feel disadvantaged in the normal competition for federal research awards and turn to Congress for a remedy. Representatives and senators, eager to aid constituents and convinced by the arguments of scientists and others that research investments are vital to future economic growth, intervene in the allocation of federal funds, guaranteeing favorable treatment for well-connected, if not yet distinguished, universities. Political review displaces peer review in research support decisions.

Politicians also have come to believe, perhaps too well, in the benefits of university research. They find it difficult to contain their enthusiasm for research investments. They want their constituents and their causes to gain a share of the expected fruits. If investments in science conquer diseases, let the work focus first on the diseases constituents champion. If investments in science bring economic benefits to the United States, then surely such investment in Texas will help the Texas economy. Unprotected by national security considerations, science is vulnerable to all the urges of democracy.[13]

Populism is surely one of those urges. Another $10 million grant to the Scranton area reveals the resentments that now have to be accommodated. This one was to the University of Scranton for a Center for Technology and Applied Research, no doubt intended to help the area's transition to full participation in the global economy. The measure was inserted into the 1991 defense appropriation bill by John P. Murtha of Pennsylvania, chairman of the House defense appropriations subcommittee, and found a Senate champion in Daniel K. Inouye of Hawaii, chairman of the Senate's counterpart subcommittee. Challenged in floor debate by advocates of competition in research awards, Senator Inouye asked, "How can Scranton compete with MIT, with all its facilities and Nobel Laureates? This is the eternal question of the 'haves' and 'havenots,' the rich and the poor. Do we make a big university bigger, and a small university smaller? Or do we provide the family farmer, the sons and daughters of coal miners and farmers an equal break?"[14] The University of Scranton received its earmarked award. The Cold War made MIT bigger. Peace, it seems, will give the University of Scranton, Marywood College, and many other schools their chance to grow.

Notes

1. The great industrial capacity of the United States and the overextension of the Axis powers won the war for the Allies.

2. The general, of course, meant there are no villains on the order of a Hitler or a Stalin threatening the United States, not that the United States had abandoned its interest in maintaining a strong military. Nevertheless, without a demon it will be difficult to maintain very sizable standing forces for very long, given the historical record.

3. For a history of the post–World War II funding patterns, see Smith (1990) and also note M. Schrage, "Blurring the line between funding science and funding economic growth," *Washington Post* (October 5, 1990):F3; and W. Lepkowski, "NSF gears up for stricter oversight of academic research support," *Chemical and Engineering News* (January 6, 1992):9–14.

4. For an argument that the military advocated and controlled the effort, see Forman (1987) and also note Geiger (1992) and Leslie (1993).

5. *Congressional Record*, 80th Cong., 1st sess., May 16, 1947, Senate 43, part 4:5425.

6. Suspicious we usually are. See Kelman (1990).

7. Among the many articles on these subjects, see Moore (1991); Spencer (1990); C. Cordes, "Federal limits on overhead rates in some programs alarm colleges," *Chronicle of Higher Education* (November 20, 1991):A33; R. M. Rosenzweig, "The debate over indirect costs raises fundamental policy issues," *Chronicle of Higher Education* (March 6, 1991):A40; and A. Pasztor, "U.S. challenge of MIT costs imperils grants," *Wall Street Journal* (January 3, 1992):B1.

8. Also see C. Cordes and J. Goodman, "Congress earmarked a record $684

million for noncompetitive projects on campuses," *Chronicle of Higher Education* (April 15, 1992):1; and C. Cordes and K. McCarron, "Academe gets $7.63 million in year from congressional pork barrel," *Chronicle of Higher Education* (June 16, 1993):A21.

9. See C. Cordes and J. Goodman, "College projects that received congressional earmarks," *Chronicle of Higher Education* (April 15, 1992):A31–A36.

10. The states feel the same pressures, as reported by W. Celis III, "Challenging a tradition of unequal universities," *New York Times* (March 18, 1992):B7.

11. The following is drawn in part from Sapolsky (1990: 127–129).

12. Also see S. Bard, "Scientists fear disease-specific lobbying hinders equitable division of U.S. funds," *Chronicle of Higher Education* (January 15, 1992):A26.

13. For an analysis arguing that democratization of support is beneficial for science, see Drew (1985). Shapley and Roy (1985) argue for a shift away from basic research.

14. Quoted in J. Long, "Pork-barrel R&D funding assailed but upheld," *Chemical and Engineering News* (November 5, 1990):5–6.

Bibliography

AAAS. 1992. *AAAS Report XVII: Research and Development FY 1993.* Intersociety Working Group. Washington, DC: AAAS Press.

Browning, G. 1992. Colleges at the trough. *National Journal* (March 7):565.

Bush, V. 1970. *Pieces of the Action.* New York: William Morrow.

Carter, L. 1966. Office of Naval Research: 20 years bring changes. *Science* 153:397–400.

Culliton, B. J. 1988. Politics of the heart. *Science* 241:283.

Drew, D. E. 1985. *Strengthening Academic Science.* New York: Praeger.

Forman, P. 1987. Behind quantum electronics: National security as basis for physical research in the United States 1940–1960. *Historical Studies in the Physical and Biological Sciences* 18(part 1):149–229.

Geiger, R. L. 1992. Science, universities, and national defense, 1945–1970. *Osiris* 7(2nd series):26–48.

Ginzberg, E., and Dutka, A. B. 1989. *The Financing of Biomedical Research.* Baltimore: Johns Hopkins University Press.

Gleckman, H. 1991. How the budget deal has choked off America's choices. *Business Week* (September 16):36.

Hager, G. 1992. Budget rules under fire. *CQ Weekly* (February 1):224.

Institute of Medicine (IOM). 1984. *Responding to Needs and Scientific Opportunity: The Organizational Structure of the National Institutes of Health.* Washington, DC: National Academy Press.

Jewett, F. B. 1971. The future of scientific research in the postwar world. In J. C. Birnbaum, ed., *Science in America*, 398–413. New York: Neale Watson.

Jones, L. R. 1992. The Pentagon squeeze. *Government Executive* (February):21–29.

Kelman, S. 1990. *Procurement and Public Management: The Fear of Discretion and the Quality of Government Performance.* Washington, DC: American Enterprise Institute.

Kevles, D. 1988. American science. In N. O. Hatch, ed., *The Professions in American History*, 107–126. Notre Dame, IN: University of Notre Dame.

Kosiak, S., and Taibi, P. 1992. *Analysis of the Fiscal Year 1993 Defense Budget Request.* Washington, DC: Defense Budget Project.

Leslie, S. W. 1993. *The Cold War and American Science.* New York: Columbia University Press.

Lewis, K. 1989. *Historical U.S. Force Structure Trends: A Primer.* RAND #P-7582. Santa Monica: RAND Corporation.

Lewis, K. 1991. U.S. force structure: Post-Gulf, post–Cold War. Lecture presented at Defense and Arms Control Studies Program, Massachusetts Institute of Technology, Cambridge, MA, October.

Marshall, E. 1991. Pork: Washington's growth industry. *Science* 254:640.

Marshall, E. 1993. The cost of scientific pork keeps going up. *Science* 260:156.

Maxwell, J. H., Blumenthal, D., and Sapolsky, H. M. 1986. Obstacles to developing and using technology: The case of the artificial heart. *International Journal of Technology Assessment in Health Care* 2(3):411–424.

Moore, W. J. 1991. Stanford's humbling lobbying effort. *National Journal* (June 1):1294–1296.

National Science Board (NSB). 1991. *Science and Engineering Indicators, 1991.* 10th ed. NSB-91-1. Washington, DC: U.S. Government Printing Office.

National Science Foundation (NSF). 1989. *Geographic Patterns: R&D in the United States.* NSF 89-317. Washington, DC: National Science Foundation.

Palca, J. 1991. NIH unveils plan for women's health project. *Science* 254:792.

Rettig, R. A. 1977. *The Cancer Crusade: The Story of the National Cancer Act of 1971.* Princeton: Princeton University Press.

Rushefsky, M. E. 1986. *Making Cancer Policy.* Albany: State University of New York Press.

Sapolsky, H. 1990. *Science and the Navy: The History of the Office of Naval Research.* Princeton: Princeton University Press.

Shapley, D., and Roy, R. 1985. *Lost at the Frontier.* Philadelphia: ISI Press.

Smith, B. L. R. 1990. *American Science Policy Since World War II.* Washington, DC: Brookings Institution.

Spencer, L. 1990. University markups. *Forbes* (December 10):37–38.

U.S. Congress. 1986. *Science Support by the Department of Defense.* Science Policy Study Background Report No. 8. Task Force on Science Policy, House of Representatives, Committee on Science and Technology, 99th Cong., 2nd sess. Washington, DC: U.S. Government Printing Office.

Walsh, J. 1987. Geographic limit on research funds in bill seen as swipe at peer review. *Science* 238:1506.

Warden, J. A., III. 1989. The air strategy for a changing world. Lecture presented at Defense and Arms Control Studies Program, Massachusetts Institute of Technology, Cambridge, MA, May.

9

Indirect Costs and the Government-University Partnership

Peter Likins and Albert H. Teich

The agreement after World War II that the United States would support basic research largely through its private and state-financed universities triggered a debate that has continued to the present day: For precisely which costs directly or indirectly associated with research should the government pay? How should these costs be calculated? Who should monitor the accounting process?

Although this debate had been in the background of federal science policy for more than four decades, in 1990 it exploded onto the scene when, in the wake of high-profile hearings on scientific misconduct, Representative John Dingell (D-MI) called Stanford University president Donald Kennedy before his subcommittee to investigate alleged abuses in indirect cost accounting at Stanford. Kennedy was caught in a doubly troubling position: first, he needed to explain a complex issue to a congressional inquiry and a public that was more interested in ferreting out and cataloguing abuses than in reforming the system; second, he was caught between admitting that the problem was Stanford's alone, and acknowledging that the problem was widespread among major research universities.

Dingell's inquiry prompted action at many levels, from hearings by other congressional committees to the return by several universities of indirect cost payments made to them by the government. As Peter Likins and Albert H. Teich recount, the circle of science policy makers in Washington also got into the act. The problems of reforming the policies for indirect costs had been compounded by the juxtaposition of the extremely complicated issues of research financing and accounting with sensationalistic portrayals of alleged abuses. But although screaming headlines prompted irate members of Congress to propose arbitrary caps on overhead, other science policy makers were able, in an environment less public but more deliberative than most congressional hearings, to turn the catalogues of abuses into detailed, productive proposals for reform.

The burden of criticizing indirect costs did not fall entirely on politicians. Many researchers themselves often view overhead charges as a kind of tax or payoff they must make to administrators in order to get the grant money they have earned, rather than as the legitimate price for the logistical support of their universities. From the perspective of the researcher, there seems to be a trade-off between direct and indirect costs, so the lower indirect costs are, the more money there will be for equipment, materials, and students—the important stuff.

Beneath the controversies, however, were the details of accounting for indirect costs. In this chapter, Likins and Teich explain the fundamental issues of indirect cost accounting. Given the complexity of indirect cost accounting and its politics, it might be more surprising that the system has worked than that is has become unhinged. Yet it is clear that the older, informal, and sometimes cozy relationships between universities and the federal agencies that fund and oversee their research are finished. They are being replaced with explicit rules, forged in the glare of public debate, that provide both incentives for and limits on universities. Yet even this new contract will remain fragile because sharp accounting practices, or even bureaucratic errors, that include questionable items in federally financed accounts will carry a cost in terms of public ridicule and political reaction heavier than the actual dollar amount. [Eds.]

Introduction

The headline in *Time* magazine's education section of March 18, 1991, was certain to grab readers' attention: "Scandal in the Laboratories."

The dictionary defines scandal as "(1) a disgraceful or discreditable action," or "(2) an offense caused by a fault or misdeed." In these terms, there is no evidence of scandal in any university research laboratory revealed by the *Time* article or any of the similar stories stimulated by congressional investigations of federally sponsored research at Stanford University. However, scandal is also defined as "(3) damage to reputation; public disgrace," and

"(4) defamatory talk; malicious gossip." In these terms we certainly do have scandal in the university research laboratories of our nation. The problem is serious because public confidence in our universities is at stake. The media have feasted on the story, but they have shed little light on issues of great importance and genuine complexity. We must try to do better.

As a first defining step, the locus of the problem must be shifted from university research laboratories to the administrative systems, policies and practices that have evolved over the past forty years to govern the relationships between universities and the research funding agencies of the federal government. This is an esoteric subject, not well suited to sound-bites on television news magazines, but it does appear that there are real problems in this area.

Neither of us is an expert, despite many years of closely related experience, but perhaps we can offer some insights. We approach the subject first as taxpayers reacting to the media coverage of this story, and in that context we look at some of the background, the causes, and the consequences of the current controversy. Finally, we examine the problem of indirect costs from the point of view of the academic community.

Amid the cries of indignation we can discern two quite distinct questions emerging: (1) why is the government spending billions of dollars each year for university overhead when that money is needed for research and education? and (2), why is federal money used for university yachts, floral arrangements, and fine silver?

The Government-University Partnership

After World War II, the United States made a rather remarkable national policy decision: it decided that the federal government should invest heavily in scientific research, and that most federally sponsored research should be done in the nation's universities. This policy was unique in world history, and it has helped to elevate scientific research in the United States to the highest standards on the globe. Other nations (most notably Russia, but also Japan) separate teaching institutions (the universities) from research institutions (government or industry laboratories), and lose the opportunity for synergy that characterizes our system. Largely as a result of this policy, U.S. universities have become intellectual beacons throughout the world, attracting graduate students, visiting scholars and permanent faculty from the international community.

U.S. colleges and universities now spend more than ten billion dollars each year pursuing research projects funded by the federal government. Not just a casual hobby performed by individual professors engaged to teach young people, this scale of operation requires special facilities and personnel. These federal research dollars pay not only professors, technicians, secretaries, and graduate students, but also librarians, accountants, and janitors who serve not only research projects but also every other purpose of the university. Not only do universities build special laboratory buildings for research, but they also build larger libraries and power plants to serve the research enterprise, together with other university activities.

From the very beginning of the unique partnership between the federal government and the research universities, one bone of contention has been the issue of whether the government should pay the full cost of the research projects. From the standpoint of a private university, the alternative is to increase the students' tuition bills, or to rely upon past or present philanthropy, both seemingly unacceptable. From the standpoint of a public university, the alternative is cost sharing by the state taxpayer. Historically this has seemed an acceptable practice in most states, although periods of austerity have often made it problematical.

The issue hinges largely on one's perspective on research as either an investment in the future, in which government is the essential funding agent and universities provide a service to the nation, or as a self-interested pursuit of the academic world for which government is providing a subsidy. Shifts in federal policy toward academic research over the years have tended to reflect the changing balance between policy makers holding these two perspectives. Our own view is closer to the former, and we would argue that, as a matter of principle, the federal government should pay the full cost of federally sponsored research at all of our universities.

In practice, however, it has proven difficult to define the full costs of research in universally acceptable ways. In order to resolve conflicts of interpretation, formal rules have been promulgated by the federal government, usually after public hearings to air differences. These rules, articulated principally in Circular A-21 of the Office of Management and Budget (OMB), are extremely complex, but their rudiments can be explained here.

Research costs may be broken down into two categories: direct costs and indirect costs. Any expenditures that can be attributed directly to a specific research project are called "direct costs." Those costs of performing sponsored research that cannot readily be attributed directly to a specific research project are called "indirect costs." Both kinds of costs are real and must be attributed to sponsored research with evidence that satisfies a government audit. Specific rulings indicate which costs are allowable in each category and which are not, and the rules of evidence are well defined.

One could argue about the rules that define the allowability of direct costs, but in practice definitions have not been a problem. Even if the research project produces a cure for a deadly disease, no one is tempted to bill the contract for the celebration that follows. Auditors do not question the costs of employing graduate students, although they realize that the students are advancing their education as they do their jobs. Although there are rare instances of financial improprieties in grant or contract management, there are no vexing public questions about direct costs. Rather, public debate focuses on indirect costs, also known as overhead; for even if all agree that research contracts should pay their full share of the costs of the libraries or the administration, there can still be arguments about what that share should be, and what elements of the total expenditure base should be disallowed. These questions can be debated for decades, and indeed they have been.

Four Decades of Indirect Costs

The rules under which the federal government reimburses universities for the indirect costs of research stem from contracting practices developed during World War II. These practices were based on the principle of "no gain, no loss" (that is, full cost recovery). During the war the principle was applied somewhat expediently as a uniform indirect cost rate of 50 percent of salaries and wages (NAS 1983: 220). When, after the war, the Office of Naval Research (ONR) became the chief supporter of academic research, it put in place a more formal set of rules for determining research costs. Nevertheless, like many practices in those more innocent days, ONR's rules sacrificed some rigor for the sake of simplicity.

As federal funding of research grew in the 1950s, the National Science Foundation (NSF) and the National Institutes of Health (NIH), which relied principally on grants rather than contracts as ONR had, took on more prominent roles. Academic scientists generally preferred grants to contracts, since they saw the former as gifts made for the purpose of carrying out research and the latter as procurement instruments requiring a clearly defined quid pro quo (Danhof 1968: 323). Use of the grant mechanism, however, undermined the concept of full cost recovery because, under a grant, the university was presumably accepting support for a project that it wanted to carry out. Thus, NIH's parent agency, the Department of Health, Education, and Welfare (HEW, the predecessor to the Department of Health and Human Services, HHS), set a limit on indirect costs for grants. Initially, this limit was 8 percent; in 1955 it was raised to 15 percent. When, in 1958, HEW proposed a 25 percent ceiling, Congress balked and subsequently set a statutory limit of 15 percent (NAS 1983: 221; GAO 1992: 12). This limit was raised to 20 percent in 1963, at which time it was applied to all other agencies. The statutory 20 percent ceiling was removed in 1966, but mandatory cost-sharing was instituted in its place through a provision in a HEW appropriations act that read: "None of the funds provided herein shall be used to pay any recipient of a grant for the conduct of a research project an amount equal to as much as the entire cost of such project" (GAO 1992: 13). The cost-sharing requirement was removed from the Department of Defense's (DOD) appropriation act in 1969, but remained in place for HHS until 1986.

By the mid-1950s, Congress and agency officials had also begun to demand more formal practices for administration of federal research funds, practices based on sound accounting principles. These demands led to the formation of a Government Interagency Committee whose deliberations (in which a group of university representatives participated) resulted in the issuance of Bureau of the Budget (BOB, now OMB) Circular A-21. This famous document, issued in 1958, laid out the rules for calculating costs applicable to both grants and contracts at colleges and universities. It defined what was allowable and what was not, and it established the notion of indirect cost "pools," discussed below. At the same time, it provided a certain amount of flexibility, allowing for the varying circumstances of different institutions. Circular A-21 has been the subject of a great deal of discussion and debate over the years. Prior to the recent flap, it had been revised at least eight times.

After Congress lifted statutory ceilings, indirect cost rates began to rise. Data from NIH indicate that indirect costs as a percentage of the total cost of a research project grant increased from an average of 15 percent in 1966 (when the limit was removed) to nearly 30 percent in 1981 (NAS 1983: 138). While some took this rise as evidence of uncontrolled growth in costs, it could also be seen as the phasing in of a more realistic cost structure, or more complete recovery of costs incurred, once the arbitrary constraints were removed. The experience of other federal agencies, subject to different kinds of legislative restrictions, differs from NIH, but overall there is little doubt that the indirect costs of university research funded by the federal government have risen faster than direct costs during the past two and a half decades. Some analysts blame increases in administrative costs for the growth in overhead. The consensus among university administrators, however, is that most of the rise is due to the expense of building new research facilities and renovating old ones—costs that in the early postwar years were borne by dedicated federal programs but that more recently have been shouldered by the universities.

Federal officials concerned about getting the most out of their university research dollars often find allies on the faculty. Many faculty members regard indirect costs as a "tax" imposed by the university on "their" research grants. At a 50 percent overhead rate, a $150,000 grant yields the researcher only $100,000 in funds that he or she can spend on equipment, graduate assistants, postdoctoral fellows, travel, summer salary, and the like. It takes a fairly sophisticated academic researcher to recognize the fact that overhead is a real and necessary component of the university's costs and not just another way for the administration to take unfair advantage of its faculty.

Too Much Money for Overhead?

Is the cost of overhead too high? Are the universities charging the federal government more than the true costs of doing sponsored research? If the indirect cost rate (ICR) at Lehigh University, for example, is 61 percent of the "modified total direct costs" (which are direct costs reduced by certain items, such as capital equipment), is that number too high or too low? If the ICR rate at Stanford is 74 percent and at UCLA 48 percent, is there something wrong somewhere? If defense contractors charge overhead rates above 100 percent, should universities increase their rates?

These quantitative questions simply cannot be answered at the level of abstract principle; the truth is in the details. The only principle that has been advanced here is that of full recovery by universities of the costs of their performance of sponsored research. That principle does not produce the same rate for all universities, and it certainly says nothing about defense contractors. According to that principle, the universities should collectively receive ten billion dollars if they spend ten billion dollars conducting research sponsored by the federal government. The appropriate share of the ten billion to be booked as indirect costs is an important issue for research scientists and university treasurers and government auditors, but it cannot be determined by watching television or reading *Time* magazine.

The Truth Is in the Details

What about the second question: Why is federal money used for university yachts, floral arrangements, and fine silver? Such questions are embarrassing, and the glare of television lights in combination with the glare of inflamed congressional interrogators can make it difficult to respond in a manner that restores confidence. Perhaps there are no acceptable answers. One must wonder how universities got in such a pickle in the first place. Is there widespread corruption in the system, or is it simply a matter of arcane procedures, poor judgment, and misunderstanding?

In order to answer these questions, it is necessary to understand the system—to recognize and separate several distinct sources of potential embarrassment relating to indirect cost charges to the federal government: (1) errors in executing accounting procedures, such as the incorrect classification of expenses; (2) the "pooling" concept, which permits costs attributable to research to be calculated as a governmentally approved percentage of all costs in a large pool of expenditures with a variety of purposes (the library pool is the familiar example);(3) memoranda of understanding (MOUs), which are official government agreements with individual universities permitting departures from general standards; (4) insensitivity in the management culture of both research universities and federal agencies, leading to reliance on "legality" when "propriety" should prevail. The following discussion explores each of these domains.

Accounting Systems

Accounting systems can always be devised to reduce the likelihood of error, but spending money in this way eventually yields diminishing returns, and no system will ever be perfect. Recording the depreciation on Stanford's 72-foot luxury yacht Victoria as a research expense seems bizarre when you see the yacht on *20/20* or in *Time* magazine, but a data entry clerk familiar mainly with the university's research vessels might be forgiven for such an error in the context of the three million other entries logged into the computer that year.

Cost Pools

The pooling of indirect costs is a reasonable and even necessary practice, but it causes real problems when the time comes to explain its consequences to a skeptical public. In the instance of the library pool, for example, some library materials are acquired strictly for research use, some for educational or even recreational purposes, and other acquisitions serve both research and teaching purposes. One approach to defining the library's indirect costs of research would be to review every acquisition every year, and classify each item individually. Because that approach would be prohibitively time-consuming and expensive, a more reasonable process has been devised. On the basis of periodic surveys it has been determined that a specific percentage of each university library's costs should be billed to sponsored research as an indirect cost. At Lehigh, these surveys produce a ratio of 15 percent; at Stanford they produce 25 percent (that seems to be a reasonable result because Stanford is a more research-intensive institution). The library cost pooling system seems to work well, but it could create a problem if its implications are challenged.

How should a university president respond if a reporter were to ask, "Is it true that you charged the federal government for your subscription to *Sports Illustrated* and called it research overhead?" Can the president reply, "Well, we charged only 15 percent of *Sports Illustrated*, and we also charged only 15 percent of *The Journal of the Astronautical Sciences*"? How would that answer play in Peoria? Or in Washington, D.C., where these rules were made?

Much more sensitive than library costs are some of the other pools approved by OMB Circular A-21. The eight pools used currently are listed without elaboration, just to illustrate the process: general administration and general expense (G&A); sponsored project administration; plant operations and maintenance; library expenses; departmental administration expenses; depreciation or use allowance; student administration and services; and interest on debt-financing of research facilities.

In each of these categories one can find the analogy of *Sports Illustrated*, but the most provocative pool is the most general, G&A. All kinds of things fall into this category, including the president's house. At Lehigh the currently approved percentage of G&A attributed to research is 8.4 percent. More research-intensive universities have higher rates; Stanford, for example, attributes 23 percent. Any cost anywhere in the university that falls into the G&A category must meet the twin tests of legality and propriety. The distinction between legality and propriety is sometimes vague; in fact costs are not legally allowable unless they are "reasonable," and that word appears to have different meanings in different campus cultures and different government agencies.

Memoranda of Understanding

The objective of MOUs, like cost pools, is to simplify accounting and documentation. The details of the controversial MOUs negotiated at Stanford over the years are not widely known. Before congressional investigations focused attention on them, there were about ninety MOUs in effect at Stanford. There are none at Lehigh and no more than thirteen at any other university. Most of the government's challenges to Stanford's indirect cost recovery stem from MOUs that were approved by ONR, apparently without adequate verification. It was once hoped that the agreements worked out at Stanford could be instituted more broadly. Obviously the lessons learned are not those anticipated.

Consideration of one MOU at Stanford illustrates the problem. Normally the governing rules require that all of the "subsistence costs" booked to the G&A and departmental administration pools be scanned (by eye or computer) to remove any "cntcrtainment" costs before they enter the pools, because "entertainment" costs are disallowed. If a professor at a research conference has an alcoholic beverage with his or her meal, that is classified as "entertainment" and disallowed from the professor's "subsistence" expense claims when the travel expense voucher is submitted. An MOU negotiated by Stanford in 1979 spares the university this scanning task, and simply reduces all "subsistence" costs automatically by 20 percent. This reasonable compromise broke down in application, however, when certain "entertainment" expenses at Stanford were charged not to "subsistence" but to other accounts not covered by the MOU. The 20 percent reduction in the MOU was therefore not applied, and the unallowable costs remained in the pool. This is the kind of error that leads to headlines: "Tax Dollars Pay for Stanford booze."

Management Culture

The charge of insensitivity in the management culture of universities to the perceptions of taxpayers is the most difficult to deal with, and perhaps the most disturbing, because it reaches to the core of the institutions, both academic and governmental. The specific issues that catch the public eye are almost always the result of errors or bad decisions deep within the bowels of a bureaucracy, but these actions occur in a cultural context that is defined by the traditions of the institution and the philosophy of its leadership. This lesson has been driven home to every university president in the United States, and to the heads of government agencies as well. Lessons learned painfully are long remembered.

The Deficit, Watergate, and John Dingell

Indirect costs have been an irritant in government-university relations for at least four decades. Never before, however, has the issue reached the current level of public visibility and political contentiousness. Why has the recent situation provoked such a storm? How has the academic community reacted? How is the affair likely to play out?

Several elements have combined to drive the indirect cost issue from the inside pages of specialized journals to the front pages of national newspapers and the trailers of television news magazines. Among them are the federal deficit, Watergate, and Representative John Dingell.

The deficit is a factor not just because federal funding for academic research represents a substantial sum of money in its own right—which it does—but also because it has been one of the few areas of growth in the otherwise bleak budget climate of the past decade. With discretionary spending squeezed in a vise year after year, the ten billion federal dollars that universities receive for their research programs are a tempting target for budget cutters, waste watchers, and anyone with an underfunded program.

As the fiscal environment has grown more cutthroat, so, too, has the nation's ethical climate changed. Emanating largely from Watergate, the widely publicized misdeeds of prominent people in government, religion, sports, and high finance have sensitized the press and the public to even the hint of scandal in every other corner of the establishment. Academia has been far from immune. The string of revelations of scientific fraud that began in the mid-1970s (see Woolf, this volume) provided a peg for disclosures about abuses of the indirect cost system—more evidence of corruption in the ivory tower.

The common thread in drawing attention to research fraud and to indirect costs was, of course, Representative John Dingell (D-MI), chairman of the House Committee on Energy and Commerce and of its Subcommittee on Investigations and Oversight. Dingell has made headlines for years by uncovering waste, fraud, and abuse in defense contracting—outrageously priced monkey wrenches, coffee makers and toilet seats. In the late 1980s, Dingell turned his attention to universities and one of their primary patrons, NIH, over which his committee has legislative jurisdiction. In the words of one observer, Dingell's focus on the Stanford yacht gave the arcane business of indirect costs "sex appeal" (Palca 1991).

Raising the Stakes

In raising the visibility of the indirect cost issue, Representative Dingell also raised the stakes. The congressional hearings and media attention created the aura of a significant problem and generated the expectation that substantial rather than cosmetic changes were needed. Discussions of the need to revise OMB Circular A-21 yet again—discussions that have never gone away but are occasionally dormant—quickly jumped up several notches on the agenda of science policy.

In the workings of Washington, once the dust begins to settle on issues of this type, they tend to be resolved through negotiations among a relatively small number of specialists from congressional staffs, federal agencies (including, in this case, OMB and the White House Office of Science and Technology Policy [OSTP]), and organizations representing the affected parties (such as the Association of American Universities [AAU]). Public concerns may set the framework for the negotiations, but the action takes place behind the scenes and with more cooperation among the participants than is generally understood.

As the indirect cost saga unfolded, the Stanford exposé was soon followed by intimations of similar abuses on other major campuses, HHS and General Accounting Office audits, and a variety of voluntary withdrawals of "inappropriate" expenses by other major research universities, including MIT. All of this created a general feeling that the universities were going to have to give in somehow on the issue. The question was not "if" but "where" and "how much."

A number of closely linked organizations headquartered in and around Washington's Dupont Circle represent the interests of higher education in matters such as this. Led by the AAU, the Association of American Medical Colleges (AAMC), the National Association of State Universities and Land Grant Colleges (NASULGC), and the Council on Government Relations (COGR), these organizations mobilized to work out the best deal they could under the circumstances. As it happened, when the Stanford situation arose, OMB was already in the midst of discussions with AAU about implementing a number of indirect cost reforms suggested by an AAU task force in 1988. These discussions were rapidly pushed aside by the new parameters.

Congressional attention focused on separating the administrative and facilities components of indirect costs and capping the administrative component. Staff from Representative Dingell's committee and its Subcommittee on Health and the Environment, chaired by Henry Waxman (D-CA), began to consider legislative options. Facing the prospect of legislation, OMB chose to preempt, and in October 1991 it issued a revision of A-21 which included a limit of 26 percent on the administrative element of indirect costs and a new, somewhat more restrictive list of unallowable expenses. While the cap was costly to a number of major institutions, the general reaction of university representatives was, "It could have been much worse" (Hamilton 1991: 788).

The new rules, which most observers saw as a temporary fix, performed several functions. They eased political pressures to respond to the revelations, while averting more draconian moves which might have further damaged the financial condition of academic science. At the same time, they preserved some flexibility for more comprehensive reforms in the future, because they came in the form of administrative action rather than legislation. Finally, they shifted the locus of action from Capitol Hill to the executive branch, where it could proceed in a less volatile (and perhaps less public) atmosphere.

In 1992, a joint working group of OSTP and OMB staff members, meeting periodically with a group of university representatives, prepared yet another revision of A-21. In HHS, a separate task force conducted a departmental-level study of the indirect cost problem. On the outside, an ad hoc collection of bench scientists and university administrators, cochaired by a Nobel laureate microbiologist and the president of a major university, and calling itself the "New Delegation for Biomedical Research," produced its own set of recommendations for dealing with the issue.

Rather than identify radical, comprehensive reforms, all of these groups reaffirmed the basic structure of the current indirect cost system and sought to make important but incremental improvements. The discussions seemed to converge on a number of key points: creating incentives for universities to keep overhead down, simplifying the rules, and making their application more consistent (Hamilton 1992). To limit the growth of indirect costs, universities would be asked to fix their rates for three to five years in advance. If actual rates turned out to be lower, they could pocket the difference; if they were higher, the universities would be responsible for the shortfall. To simplify the often complex accounting required of universities, a threshold rate was proposed for the administrative component of indirect costs. Universities could claim the threshold rate without detailed backup, the way individuals can claim the standard deduction on their income taxes (Hamilton 1992). To address inconsistencies in the way different universities handle research costs, the discussions proposed standard methods for calculating indirect cost rates and for treating various categories of expenditure.

Most of these suggestions were incorporated into the proposed revision of Circular A-21, which finally appeared for public comment in the *Federal Register* on December 9, 1992 (Anderson 1992). Threats to set an upper limit on overall indirect cost rates or to set a standard rate for all universities did not materialize, to the relief of most of the academic world. However, the limit of 26 percent on administrative costs was retained, and the threshold rate was set at 24 percent. With such a small margin between the threshold and the cap, observers speculated that most universities would find it more advantageous to accept the 24 percent threshold rate rather than pay for extra accounting services to gain an additional 2 percent. Thus, 24 percent could become the effective cap. In addition, by including certain other costs within the administrative category, the new rules reduce further overhead recovery for those institutions whose administrative costs are already above the cap. A new prohibition on including tuition remission for graduate students employed on research projects will have an especially adverse effect on such research-intensive universities as MIT, Caltech, Columbia, and Stanford.

The problem of how to explain and deal with the differences in indirect cost rates among institutions was tabled for further study. Communicating the notion that it is fair and reasonable for consistent rules to yield different outcomes at different institutions is something that has so far eluded both policy makers and the leaders of the academic world. Also left in the hands of the Clinton administration is the thorny problem of facilities costs. The need to expand and upgrade academic facilities continues, and the initiatives approved during the course of the fiscal year 1994 budget process, while a step in the right direction, will barely begin to address this problem.[1]

Although the changes to A-21 were expected to put the issue of indirect costs on the back burner, budgetary pressures kept it simmering in the succeeding months. A bill sponsored by Senators Orrin Hatch (R-UT) and Bob Dole (R-KS) sought to pay for increased crime-fighting efforts with a 10 percent across-the-board cut in university overhead. While this measure did not get far, a more serious issue was raised by the inclusion of a 50 percent cap on indirect costs in a piece of deficit-reduction legislation introduced late in the 1993 congressional session. Though the bill, known as the Penny-Kasich Deficit Reduction Plan after its cosponsors in the House of Representatives, Timothy Penny (D-MN) and John Kasich (R-OH), was ultimately defeated, it came close enough to passage to cause serious concern among academic administrators and their Washington representatives. Penny-Kasich offered no analysis to back up its proposed cut; it simply treated the indirect costs of research at academic institutions as an area of federal spending that could be arbitrarily reduced in a difficult budgetary environment.

How could the same Congress that supported an 8 percent increase in the NSF budget display such insensitivity to the realities of university finances? Think of it as the return of the Stanford yacht.

Source Note

This paper is a revised and expanded version of Likins (1991).

Note

1. The Clinton administration had requested an increase of $5 million in NSF's $50 million "Academic Research Facilities and Instrumentation" program. Congress added $55 million to the request, bringing the effort to $110 million, and renamed it "Academic Research Infrastructure." See Teich et al. (1993).

Bibliography

Anderson, C. 1992. Universities discover that simplicity has its price. *Science* 258:1874.

Danhof, C. H. 1968. *Government Contracting and Technological Change.* Washington, DC: The Brookings Institution.

General Accounting Office (GAO). 1992. *Federal Research: System for Reimbursing Universities' Indirect Costs Should Be Reevaluated.* GAO/RCED-92-203. Washington, DC: GAO.

Hamilton, D. 1991. Indirect costs: Round II. *Science* 254:788.

Hamilton, D. 1992. Indirect costs: A consensus on reform begins to take shape. *Science* 258:737.

Likins, P. 1991. Scandal in the laboratories? In M. O. Meredith, S. D. Nelson, and A. H. Teich, eds., *AAAS Science and Technology Policy Yearbook, 1991*, 373–381. Washington, DC: AAAS Press.

National Academy of Sciences (NAS). 1983. *Strengthening the Government-University Partnership in Science.* Ad Hoc Committee on Government-University Relationships in Support of Science. Committee on Science, Engineering and Public Policy. Washington, DC: National Academy Press.

Palca, J. 1991. Indirect costs: The gathering storm. *Science* 252:636–638.

Teich, A. H., Nelson, S. D., Gramp, K. M., and Gehman, P. R., Jr. 1993. *Congressional Action on Research and Development in the FY 1994 Budget*. Washington, DC, American Association for the Advancement of Science, 45.

10

Research in U.S. Universities in a Technologically Competitive World

Eugene B. Skolnikoff

Until the rise of Fascism in Europe began to drive European scientists to the United States, few American scientific careers were complete without a sojourn with colleagues across the Atlantic. The openness of European universities to these visitors was a boon to American scientists, who returned to the United States with new knowledge, tools, and even ideas about the organization of laboratories and universities.

Since World War II, the situation has been reversed. The United States became the principal destination for the largest number of traveling students and scientists. In an era when the human capital and intellectual property developed in universities are increasingly important to a nation's economic position in the global economy—and when in this country public-private partnerships in the university are a primary means for this development—the flow of foreign scientists and engineers and their activities in universities becomes an important question.

In this chapter, Eugene B. Skolnikoff discusses some of the tensions between the national and international roles of a university. Skolnikoff draws extensively on his experience as the chairperson of the Faculty Study Group on the International Relationships of MIT. This group explored and responded to questions—in part raised by the late Representative Ted Weiss (D-NY) in hearings held in 1988—about potential conflicts between universities as members of an international community of knowledge, and universities as a focus of research for technological innovation and national advantage.

Skolnikoff identifies a number of legitimate questions for the universities to confront in a new environment characterized by greater international economic integration and a more level international scientific playing field.

Economics, demography, and technology challenge the old role of the open university, making visiting researchers, foreign students, and investment by foreign corporations potential threats.

Skolnikoff argues that while these potential threats require attention by universities, they do not currently warrant extreme action. He does, however, identify several areas that require continued attention and recommends that universities consider the creation of a forum in which they might be able to resolve the tension between their national and international roles. Without such a forum, discussion of this tension could become enmeshed in a political debate about foreign trade, and this in itself could threaten the integrity of the fragile contract between science and society. [Eds.]

Introduction

The relatively poor performance of the U.S. economy in high technology competition, especially with Japan, has stimulated criticism of the policies of research universities toward foreign corporations and students. Questions have been raised asking whether those universities, supported by public taxpayer funds, are providing knowledge and training that is used by foreign countries to the economic disadvantage of the United States. These questions, sometimes couched more stridently as accusations of misbehavior, are raised primarily by a few in Congress, echoed by some in the executive branch and industry, and often amplified in the media. Their focus is on many different aspects of university activities, for example the training of foreign students, accessibility of research laboratories to visitors from other countries, conduct of industrial liaison programs that provide foreign companies with organized access to research, support of research by foreign sponsors, and licensing of patents to foreign corporations.

The questions most frequently put forward are:

- Do foreign corporations gain early access to research paid for by U.S. taxpayers to the economic detriment of U.S. corporations?
- Does foreign support of research skew the direction of university research toward the interests of foreign corporations?
- Do gifts from and other relations with foreign corporations give them special access to and influence over research laboratories?

• Is access to university laboratories given too easily and, in effect, underpriced, constituting "cream-skimming" of a research base paid for by U.S. taxpayers?

• Do U.S. universities admit too many foreign students who later work in industry abroad, using the knowledge gained from their university training to compete with U.S. corporations?

In the current economic climate these are legitimate questions. In effect, they can be aggregated to ask whether the traditional openness of the country's research universities in their international interactions continues to be appropriate when it appears that other nations are more adept than the United States at using the knowledge gained for economic purposes. Or, to put it in a more general way, are the national and international responsibilities of the research university now in conflict?

The Changed Environment of U.S. Research Universities

Since World War II, research universities in the United States have provided education and scholarship of a quality that is the envy of the rest of the world. They have given generations of students advanced knowledge and skills that have benefited industry, government, and the universities themselves; and they have taken a leadership role in scientific and technological endeavors that have served the nation's economic, security and social purposes extremely well. Today, however, the universities face new challenges stemming from the dramatically changed social, technological, economic, and political setting in which they are embedded.

Perhaps the most significant change in the research setting is the new level of international integration of societies and economies. That integration is characterized by the internationalization of industry, easier and more extensive communication across borders, increased dependence within any one country on developments and decisions in others, and growth of issues and technologies with worldwide dimensions. With this higher level of interdependence has come difficult international problems in economic, environmental and security areas, and a wholly new level of interaction in science- and technology-related issues across national borders.

At the same time, the level of competence in science and technology has risen markedly in most industrialized and many developing countries, so that the postwar dominance of the United States has significantly eroded. As a result, it is now essential for the health of science and technology in the United States, as it was not in the immediate postwar years, for U.S. researchers to have access to and keep abreast of research throughout the world.[1] The nation's students and faculty must be educated to be able to engage in international exchange of knowledge, at the same time that other nations are now much better positioned to benefit from research and development done in the United States.

The apparently closer relationship than in the past between the laboratory and the commercial marketplace is another substantial shift. In the United States, support for science has been predicated in part on the likelihood of economic benefit sometime in the future. Now, shorter product cycles, the increased science-dependence of some technologies, and the entrepreneurial activity of faculty and students in many fields suggest a closer, more immediate tie between research and economic benefit. The actual relationship is complex and varies among fields and technologies.

Though the promise of quicker economic returns has been a spur to the support of science in the universities, it also has raised the level of concern about who has access to the research and when, and about who benefits from it. If other nations benefit more than the United States, some ask why the government should be providing support. Moreover, the growing significance of economic interests in the results of research has raised more directly the possibility of conflict of interest between the traditional values of open research and pressures for confidentiality because of potential commercial applications.

In this globally competitive, technologically dynamic economy, U.S. industry has in general not fared well and finds itself under severe challenge from foreign industry.[2] The reasons are complex and disputed, relating to management competence, time horizons for measurement of corporate performance, industrial structure, availability and cost of capital, training of the work force, government policy, and a host of other matters, in particular adeptness at translating the ideas of the laboratory into commercial products (Dertouzos et al. 1989).

Added to the overall picture is changing demographics: the pool of students born in the United States who might become engineers and scientists is shrinking, and the proportion of college-aged people who come from disadvantaged educational backgrounds is increasing. In addition, the nationwide weakness of primary and secondary education has had a significant effect on higher education in science, engineering, and management. U.S. research universities and industry are becoming increasingly dependent on students and trained workers from abroad who elect to remain in the country, especially those with graduate degrees. Fortunately, from this perspective, the quality of higher education in the United States remains attractive to students from foreign countries who, by remaining here, enhance the country's intellectual climate and research base, and help to staff its industry and universities.

Finally, the changing financial situation of U.S. universities is an important factor. The uncertainty of public resources for research at a time of increasing educational, research, and facilities costs has led universities to mount aggressive fund-raising activities. The decline in federal support for research equipment and facilities in recent years has increased the need to find resources to supplement those from public sources. For private universities in particular, the availability of resources is a matter of considerable anxiety, with little prospect that the situation will improve in the near future. The result is a continuing search for new and previously untapped sources.

The search for new sources of funds is exacerbated by apparent changes in attitudes toward universities and science that has accompanied a growing populism in the United States. The prospect that the otherwise welcome end of the Cold War may have negative financial implications for the support of university research by the federal government—and in particular on the merit-based distribution of public funds for research—has raised great concern (Sapolsky, this volume).

General Responsibilities of Research Universities

To evaluate the issues that have been raised, it is well to start from first principles about the responsibilities and missions of the research university.[3]

The research enterprises of research universities in the United States, though they have many roots, are in their present scale and configuration a product of the nation's structure, history, and policies—particularly the period after World War II when public funds expended by the federal government became the primary source of research support. In 1990, for example, federal funds accounted for 75 percent of the research budget at the Massachusetts Institute of Technology (MIT), and federal funds have contributed to a considerable portion of the physical plant and research facilities. Support from public funds is typically an even higher proportion of total research support at many other research universities. Thus, these universities can be seen as unique products of the nation that formed and nurtured them, and as integral and essential elements of the nation's education and research system. Given their origins and roles, the research universities clearly have a special responsibility to the nation to perform as well as they are able their central missions of education and research.

That goal has several implications. For one, it mandates a commitment to quality research and education, meaning that quality must be a primary criterion in the selection of faculty and research staff and in the admission of students. Compromise in that criterion on grounds of nationality, source of support, or other parochial considerations will inescapably lead to lowered intellectual capacity and performance.

The goal of maximum performance also means that in today's world, where first-class competence in science and technology is widespread, the universities must be able and willing to interact fully and openly with research activities throughout the world. The universities must maintain openness of research and education, must be active participants in international scientific and technological communities, and faculty and research staff must be able to interact freely with colleagues abroad and have ready access to research in other countries.

Such a commitment to openness and full participation in international activities is also an expression of the responsibility of the academic community to an international scientific community dedicated to the free and open exchange of ideas.

At the same time, many intellectually challenging and socially important problems are national, regional, or local in scope.

Among the most significant is the health of the U.S. economy which, though heavily influenced by developments in the world economy, depends ultimately on actions taken within the nation. The intellectual focus of research universities on science and technology, their responsibility to the nation, and their own self-interest, all dictate a vital concern with the performance of the U.S. economy. At a time when domestic productivity growth is lagging and international economic competition is intensifying, a focus on demanding problems of U.S. industry and on the effective transfer of knowledge to U.S. industry are therefore appropriately important aspects of a university's mission.

Thus, research universities have responsibilities to the nation that call for open interaction with the global science and technology community while also focusing on the problems of performance in U.S. industry. These missions need not be in conflict; in fact, they should be mutually supportive. But, as with all general statements of goals, policies designed in their pursuit often produce situations characterized by considerable ambiguity, with the emergence of apparent or actual conflict.

Conflicts between National and International Roles

Access to Research

Access to research at the university is just one, and perhaps not the most difficult, step in the commercialization of new processes or products. Of greater importance is the ability of industry to convert new ideas and information into high-quality commercial products rapidly and efficiently. *Made in America* (Dertouzos et al. 1989) and other studies have provided ample evidence that U.S. firms in a broad range of industries have been lagging behind their competitors in other countries in this key dimension of performance. The causes are varied and are not confined to industry; universities also share some of the responsibility, in particular through the education and training of those who work in and direct U.S. industry.

MIT and other universities, in cooperation with industry, have responded by developing programs designed to address facets of the problem—for example, programs focusing on manufacturing technologies and management—that were long neglected in the

United States. These and other efforts to restructure industrial patterns will have an impact only in the long term; the immediate problem remains the effectiveness of translation of knowledge to the commercial marketplace.

Given that situation, other nations that are effective in the commercialization of knowledge derived from research have undoubtedly benefited and will continue to benefit economically from access to the products of U.S. investment in science and technology. Indeed, improving their ability to do so was an explicit objective of U.S. foreign policy for many years after World War II. (The United States benefited in a similar way from the science and technology investments of European nations before the war.)

Acknowledging these benefits does not mean that knowledge flows disproportionally to foreign firms. Results of a recent survey of the faculty and research staff at MIT suggest that the preponderance of their formal and informal contacts with industry is with U.S.-based corporations. About two and a half times as many personal contacts were reported with U.S. firms than with either Western European or Japanese firms, about 40 percent more than both taken together (Westney 1992). Proximity and cultural factors also help to make the transfer of knowledge easier and more effective in these much more numerous contacts with U.S. firms than with foreign firms, other factors being equal.

Nevertheless, the fact that some foreign companies may benefit disproportionally through their advantage in commercialization of knowledge has raised the question of whether any conditions should be placed on access to this research by individuals or corporations from abroad who have not contributed to the development of the research base, but who can make more timely use of it. Some have advocated broad restrictions on access by foreigners or at least special payments to compensate for "cream-skimming."[4]

Such broad restrictions would be in conflict with the general principles of openness laid out above, and presumably should and would be rejected by universities. It is well to ask, however, whether all the routes for access to research deserve equal protection as a matter of principle or practicality. Should there be any conditions imposed on some, or special charges attached? An answer requires separate examination of the various routes for access.

Visiting Faculty, Research Scientists, and Postdoctoral Fellows

A particularly important route for learning about research is through visits to faculty and laboratories. Visitors can gain access to research results in preliminary and prepublished form and can also learn about the direction of research and about the more intangible, embedded or "know-how" aspects of processes and techniques.

The most effective transfer of knowledge usually occurs in the course of long-term visits (a semester or more) where visitors and inside researchers have equivalent knowledge and are working toward similar goals. Faculty routinely invite long-term visitors on the basis of their knowledge and expertise, both to contribute to research projects and to learn from their stay. In some fields, in fact, postdoctoral experience is considered an essential qualification for research and academic positions. Exact numbers of visitors are not typically available, though foreign visitors are usually known because of visa requirements. At MIT in academic year 1989–1990, there were some 1,250 long-term foreign visitors on campus, including professors, visiting scientists and engineers, postdoctoral fellows, research affiliates, and others, of whom fewer than 100 were paid fully or partially by foreign industrial firms.[5]

If foreign visitors, including those from industry, are completely open with regard to their knowledge and expertise, they are a valuable, and in some fields essential, resource. They make substantial contributions to U.S. research objectives, provide up-to-date information about work and goals in other countries, and in many cases stay on to fill significant positions in U.S. universities and industry.

Restrictions on the access of foreign visitors would impair a source of ideas and skills that becomes increasingly important as scientific and technological quality improves abroad. But implicit in this analysis is that openness in the exchange of knowledge should be the norm for long-term visitors to universities from foreign countries (or from the United States for that matter). Such a rule cannot be defined with precision; nor can compliance be fully assured in advance. Essentially, the host should have a realistic expectation that visitors will participate appropriately in the research underway, be willing to share their own skills and knowledge, offer seminars or other opportunities for presenting work being carried on elsewhere, and in general be full scientific colleagues without constraint.[6] There would appear to be no basis, in principle or practice, to treat specific areas of research that may be close to industrial work differently, but the condition of openness on the part of visitors becomes even more important in those cases.

Who should bear the responsibility for ensuring that the condition of openness is met? The professional judgment involved in evaluating openness dictates that it can be done only at the level of the individual faculty or laboratory, perhaps with strong urging from university administrations. Centralized responsibility would be both impractical and inimical to the research enterprise.

Occasionally, the right to send a visitor is written into a gift or other arrangement made between a company and a university. These can be problematic situations, depending on how individuals are actually nominated, and who has the final say in their choice. To satisfy the requirement of quality control and openness, acceptance of a visitor must ultimately always be at the discretion of the university, not dictated by a donor or other outside source.

The question of reciprocity in visits is often raised, with the assumption that visits from abroad ought to be balanced with "return" invitations. Most scientific relationships involve a generalized reciprocity in the flow of information, travel, exchanges, communications and visits, usually over many years. The current difficulty in assuring reciprocity arises primarily with Japan because of the asymmetry in the structure of scientific research between the United States and Japan. By and large, university research in that country holds little interest for U.S. researchers. The most interesting work is conducted not at the universities but in industry, where conditions of proprietary information and closed laboratories prevail, as they do in U.S. industry.

In this situation, reciprocity in visits may not be feasible. Anecdotal evidence, however, indicates that there are differences among fields of research; in some fields it is apparently quite easy for U.S. academics to visit and work in industrial laboratories in Japan, while in others barriers are encountered. There would seem to be little to be lost in a general injunction to seek reciprocal visits whenever practical, recognizing that the results would be uneven.

Though most of the foreign visitors at MIT were not from industry, but from universities, governments, and other organizations, there were probably more long-term visitors from foreign than from domestic industry. (Records on domestic visitors are unavoidably incomplete.) This is apparently because of a different, and lesser, valuation placed by U.S. industry on the knowledge and skills that can be gained from such extended visits. There have been a number of unsuccessful efforts by MIT faculty to induce U.S. companies to send staff for extended participation in economically promising research programs. Many of these failed efforts then proved to be of considerable interest to foreign firms.

This attitude of U.S. industry is presumably a result of an industrial culture with standard routes for advancement that give little credit for midcareer university experience, the mobility of engineers and scientists, and other factors such as pressure on industry to demonstrate short-term financial returns. It is another example of the different patterns of relationships between industry and laboratories in the United States and other technologically advanced countries.

Short-Term Visits

Short-term visits do not typically result in a significant transfer of knowledge beyond what is generally available through publications and conferences, especially if a visitor is not deeply versed in the area of interest. Such visits would not be expected to be fully equivalent in the flow of information between the participants, typically being more of a show-and-tell, along the lines of a research report. On the other hand, even short visits to a laboratory can elicit considerable detailed technical information of potential commercial value if a visitor is well-qualified, prepared, and interested in specific material. In those cases, a visit can be a two-way exchange, but only if the faculty member is prepared to make it so.

But the subject of short-term visits to research laboratories is difficult to deal with in political terms, for such visits are often overvalued as an effective means of transferring knowledge. This political sensitivity is especially marked with regard to programs designed to further the transfer of technology to industry while attracting industrial financial support. The Industrial Liaison Program (ILP) at MIT, established in 1948, is a good example to look at because it has become a lightning rod for criticism as the proportion of foreign companies in the program has increased. While other universities have established similar programs, the MIT program has received particular attention because of its size and history.

MIT established the ILP in consonance with the university's traditional emphasis on transfer of technology to the community, a mission stemming from MIT's inception as a land-grant institution in the 1860s. The inspired merger of that mission with the need for additional sources of funding produced a program after World War II that has been of considerable value to the Institute's activities. The ILP has been responsible for developing many productive relationships between the faculty and industry, and it also contributes some $3 million to MIT's budget each year (out of a total income to the ILP of approximately $8 million).

ILP charges member companies a fee, in return for which they are provided with assistance in keeping abreast of work at the Institute. This assistance includes invitations to special symposia, help in organizing visits to faculty and laboratories of interest, distribution of summaries of research underway, occasional visits of faculty to company sites, and distribution of publications. The program employs about twenty industrial liaison officers. As of March 1991, there were 245 corporate members, of which 121 were foreign, including 57 from Japan, 56 from Europe, and 8 from other countries. The ILP has an office in Tokyo that was established in 1976 specifically to expand and offer services to Japanese members. That office has also begun to provide support to other MIT activities in Japan.

The ILP facilitates access to MIT on the part of member companies but does not provide privileged access: all the information available through the ILP is equally available to nonmembers on their own initiative. Clearly, however, a company that uses the ILP has an advantage in that it will learn about research of interest earlier than it might otherwise and can obtain information and make contact with faculty more efficiently. This efficiency is the primary motivation for companies to join the program.

The faculty at MIT generally find the ILP a reasonably useful vehicle for obtaining information from industry (Westney 1992). They also appreciate the incentive system that makes a portion of the program's income available to participating faculty for research materials and professional expenditures. It cannot be said that there is always, or even usually, reciprocal exchange of knowledge during ILP-mediated visits—that is not the purpose of the program—but in fact it is often considered by the faculty to be a valuable vehicle for staying abreast of advanced industrial research. However, some ILP-mediated visits do involve carefully prepared and directed industry representatives who can gain information of value to their companies. Contacts that involve serious, substantive transfers of knowledge frequently evolve into another form of relationship, such as sponsored research or private consulting arrangements.

In evaluating the overall role of the ILP in contacts between the faculty and foreign corporations, it is important to realize that industrial contacts arranged through the ILP constitute only a small portion of all relationships between the MIT faculty and industry. In particular, relationships with U.S. companies are considerably more numerous through direct contacts than through the ILP. The survey data show that the faculty have considerably more contacts with U.S. firms outside the ILP than through the ILP (Westney 1992). Foreign companies, on the other hand, tend to have a higher level of activity in the ILP than their U.S. counterparts; they typically have greater difficulty (for cultural, linguistic and geographical reasons) than U.S. firms in developing their own direct relationships with the faculty. U.S. firms are more likely simply to telephone faculty members, even if they have not previously been in contact.

The MIT Study Group that examined the Institute's international relationships considered whether the ILP may have outlived its usefulness in the transfer of knowledge to industry and as a means of raising resources. Its prominence in MIT's relationships with foreign corporations, which has attracted much attention and criticism (sometimes using the argument that MIT is "selling" research paid for by the U.S. taxpayer), is also of concern. However, the value of the ILP to the faculty as a vehicle for staying abreast of industrial research, the program's usefulness in raising resources, and the fact that the outside criticism of the program does not accurately reflect either its actual operation or the extent of the overall relations between MIT and U.S. industry, argued strongly in its favor and gave no grounds for recommending reevaluation. Indeed, the Study Group expressed concern that U.S. firms do not participate more in the ILP.

On similar grounds, the Study Group saw no basis for establishing a limit on the proportion of foreign-based companies in the ILP or for restricting the provision of services to member companies based on nationality. Restrictions of services to a subset of companies would in any case be impractical and undesirable to attempt to administer.

The Study Group did see higher fees for foreign companies as a more appropriate means of reflecting the benefits that foreign corporations receive; this policy is already implemented in the program's fee structure.

The existence of a Tokyo office of the ILP has been criticized as providing a special advantage for Japanese companies, even though the original motivation was to redress the difficulties that distance and cultural differences pose for management of MIT's relationships in Japan. The Study Group instead saw the office as an opportunity to increase diffusion of knowledge about the Japanese scientific and technological community. It recommended that its role be expanded to assist members of the MIT community and non-Japanese members of the ILP to become more familiar with science and technology in Japan and to contribute to the Institute's research and education interests. The Study Group also suggested that a comparable office might be established in Europe to serve similar purposes.

Support of Research

Universities undertake sponsored research that advances their missions in education and research, fulfills the intellectual interests of the faculty, provides an enriching experience for students, and can be freely and openly reported. One of the key elements in the strength of U.S. research universities, compared with those in other countries, has been their commitment to the integration of research and education.

In recent years, universities have striven to increase their funding from industrial sources, raising the proportion of industrial funding of R&D to 7 percent in 1990 from 4.3 percent in 1981 (NSF 1990a,b). Some 5 percent of those funds came from foreign firms (GAO 1988). MIT has a higher proportion than the average, attracting some 15 percent of on-campus funding from industry, with 20 percent of that ($8.7 million or 3 percent of all on-campus funding) from foreign industry in fiscal year 1990.

Although the support of research by foreign industry is not very great, concerns have been raised as to whether foreign support diverts the research toward problems defined by foreign, rather than U.S., industry.[7] There is no doubt that there are complex interactions between the availability of funding and the setting of research objectives. But as long as (1) the faculty has autonomy in deciding what research has scientific and technological merit, (2) the work has similar conditions for access as other research at the university and will be freely and openly reported, (3) all qualified students are eligible for participation whatever their country of origin and the research sponsor has no say in their selection, and (4) there are no atypical benefits of the results accruing to the sponsor, there is no evident reason why foreign research support should raise any particular concern. In fact, such support provides mutual benefit to all parties, for it helps to keep U.S. academic research abreast of the research frontiers of interest to foreign enterprises, provides information in the process about the state of knowledge abroad, and helps to maintain research and education in the United States.[8]

Faculty Consulting

In addition to meeting and working with industrial visitors on campus, faculty members often visit industrial sites. Their purposes are diverse. Some are consultants, in accordance with the typical rule allowing one day per week for private professional activities. Some have been invited because a corporation is supporting a research project, or has donated the chair held by the faculty member. Others meet with scientific colleagues to exchange information. Still others may be conducting research.

These visits have long been accepted as a natural aspect of a university appointment. They serve to stimulate professional development, provide a channel for knowledge transfer to and from industry, and give faculty members new insights into industrial problems. At MIT, preliminary survey results indicate that more than two-thirds of the consulting contacts of faculty with industry are with U.S. industry, implying that the transfer of knowledge through these visits continues to be mainly to U.S. industry.

Visits to foreign corporations, however, can also be valuable to education and research missions. For example, a 1987 survey of the MIT faculty found that the most important correlate of faculty interest in Japanese scientific and technical literature—an increasingly important source of information—was a visit to Japan (Samuels and Westney 1987).

There can be little doubt that consulting by faculty is a route for transferring knowledge; that is, after all, one of the reasons industry employs faculty consultants. In this role, the faculty act as private agents, with relationships established on the basis of their knowledge and capability, independent of their affiliation. The question can of course be raised, as it can with any private business transaction, as to whether or not larger national interests are reflected in such relationships when they involve a foreign industry. But any attempt to place restrictions on faculty consulting with foreign industry would thereby raise issues of interference with private property and enterprise.

Gifts

Gifts (for example, endowing a faculty chair) are included under access to research only because of the possibility that universities might accord special privileges to major donors through control of the appointment of an individual to a chair, through influence on the direction of research conducted under the chair, or through special relationships to that research.

Universities may have different policies with regard to the conditions under which gifts are accepted. At MIT, it is felt that gifts in support of the infrastructure of education and research are the most appropriate way for foreign beneficiaries of university research to contribute to the continued productivity of the research base. Gifts made in that spirit influence the direction of research in only the most general way, by strengthening a major area of research; donors do not set the problems or goals for MIT research. The Institute expects to develop a relationship, even a close relationship, with its major donors, including those from abroad. But even a close relationship does not involve the provision of privileged services to the donor. Chairholders may make occasional visits to the corporations that donate chairs for the purpose of reporting on research, but they do so as a courtesy and not to provide information unavailable to others. The selection of the individual for a chair is strictly a prerogative of the Institute.

In some cases, MIT has granted ILP membership for a limited period to corporations donating chairs in the hope of developing substantive relationships between faculty and corporations that are of technical interest and that might lead to sponsored research or other additional support. Those memberships either grow to become regular memberships or are terminated. In a few cases, agreements at the time of receiving major gifts have included a provision for the donor to send a visitor to a laboratory. Such visitors are required to meet the usual criteria for visitors as discussed above; in all cases, the acceptance of a proposed visitor is at the discretion of MIT, not the donor.

The same considerations apply to chairs endowed by domestic or foreign donors. As of 1991, approximately 30 of the 215 endowed chairs at the Institute had been funded by foreign-based corporations, with some 25 from Japan. The argument has been made that the large number of chairs from Japanese corporations must reflect their calculation of the value they have received from their ties to MIT. This is undoubtedly so, but it is equally true of U.S. donors. Whether the support through unrestricted gifts is an adequate recompense to the university for the value received by the donor, regardless of national origin, is a more subtle question.[9] But one can only be sure that the MIT administration seeks to maximize unrestricted gifts.

Patent Policy and Entrepreneurship

Ownership and control of international property rights that emerge from the university is of growing concern in international competitiveness. Again, the policies of different universities may vary; only those of MIT are summarized.

As part of its commitment to the transfer of technology to the larger community, MIT has a long tradition of patenting inventions made by its faculty and students and licensing them to entrepreneurs and established industry. This tradition has been reinforced in recent years by changes in federal policy that clarify ownership of patents for inventions made under federal research grants and allow university ownership as a way of stimulating innovation. As a result, MIT has given greater attention to protecting and licensing its intellectual property.

Federal guidelines mandate that licenses for inventions developed with federal support require "substantial" manufacturing of the licensed product in the United States if the product is to be sold in the United States. MIT goes beyond these guidelines by requiring that licensees of all MIT inventions, not just those that result from federal research support, be subject to the "substantial" domestic manufacturing provision. The policy is increases the prospect of successful commercialization because proximity to the inventor is so important in the process. As a bonus, the policy frees MIT from the administrative headache of determining whether or not a given invention was supported by federal funds. In fiscal year 1990, 84 percent of MIT licenses were with U.S. firms, and 16 percent were with foreign firms that have U.S. facilities.

Beyond licensing, MIT's Technology Licensing Office also endeavors to help establish new ventures based on MIT inventions, following the pattern of entrepreneurial activity that has long characterized the ethos of the Institute. Over the years, new companies started by graduates and professors from MIT have contributed substantially to the U.S. economy, especially in high-technology fields. Recent studies of Massachusetts and Silicon Valley attributed some 450,000 jobs directly to such companies as Raytheon, Digital Equipment, and Hewlett-Packard, all spin-offs of MIT faculty (Chase Manhattan 1990). The youngest firms tend to be in the most rapidly growing, high-technology fields; for example, individuals related to MIT started fifty-seven software and twenty biotechnology firms in Massachusetts between 1980 and 1990 (Bank of Boston 1989; Chase Manhattan 1990).

It is possible that other universities have been willing to establish more exclusive patent arrangements with foreign supporters of research when federal funds were not involved. Clearly, such arrangements would raise more direct questions about where the benefits of research will be realized.

Students

Students are a major channel for the transfer of knowledge from academia to industry. Graduates, whether U.S. citizens or not, who join industrial firms abroad may become factors in increasing foreign competition for U.S. firms. Similarly, U.S. firms hire both U.S. and foreign graduates, who then contribute to the competitiveness of the United States.

The United States has the largest number of foreign students of any country (408,000 in 1991), comprising about 1.5 percent of undergraduate and 12 percent of graduate enrollment (NSF 1993). In research universities, the proportions are generally higher: foreign undergraduate and graduate enrollment are about 9 percent and 33 percent, respectively, at MIT.

Foreign students at the undergraduate level at MIT are limited as a matter of policy. That is, high-quality applicants are numerous and could be admitted in much larger numbers. The origin of the rationale for 9 percent (up from 5.2 percent in 1979) is murky, but it is largely attributable to a compromise between a commitment to diversity in the student body and a desire to serve national needs by saving places for U.S. students. The pool of U.S. students applying is large enough and good enough that there is no sacrifice of quality in an arbitrary limitation on the number of foreign students admitted. Whether that applies at other schools is not known.

At the graduate level in science and engineering, research and education become essentially indistinguishable, with academic quality—based on preparation and likelihood of contributing to research—being the main criterion for admission. The 33 percent of international students at MIT, a high figure among research universities, compares with about 26 percent a decade ago. In some areas of study, however, MIT's international population is not unusually large. About 40 percent of the MIT doctorates awarded in engineering, for instance, are received by students who are not citizens, a figure that is less than the national average of more than 50 percent (NSF 1993). The distribution of international students across the graduate fields of the Institute is quite uneven. In the sciences, international students make up about 34 percent of the graduate student population in physics, 55 percent in math, but only 14 percent in biology. In engineering, international students comprise 25 percent of the graduate student population in electrical engineering and computer science (EECS), 50 percent in nuclear engineering, and 56 percent in civil engineering.

Unlike undergraduates, graduate students are admitted by a process administered by academic departments. Procedures for admission of international students vary widely. In some departments at MIT such as biology and EECS, the number of high-quality applicants from both the United States and abroad is far greater than the number of places. The graduate admissions process in these cases resembles that for the undergraduate student body in that limits (formal or informal) on admission of foreign applicants are imposed, and have been in place for many years. The proportion of international graduate students in these departments tends to be lower than the Institute average. The policies of these departments to limit foreign students are largely motivated by a desire to ensure adequate opportunities for U.S. students, without violating the criterion of quality. Whether this is valid for other universities is unknown.

Certainly, the size of the qualified pool of U.S. students in high-value fields is an issue that must be of long-range concern. It was the judgment of the biologists at MIT, for example, that there are only a handful of U.S. universities that have enough qualified U.S. applicants to be able to limit foreign student enrollment without any sacrifice of student quality.

A different issue, arising out of the concern that U.S. universities are training the nation's economic competitors, is whether foreign students should be allowed to work in research areas that are of particular economic sensitivity. MIT's policy is that once admitted, there should be no test of nationality with regard to the research on which students work.

Some of the recent curricular initiatives at MIT designed to address U.S. needs raise this issue in a new form. These initiatives, often in collaboration with U.S. industry (such as Leaders for Manufacturing, conducted jointly by the Schools of Engineering and Management), naturally target U.S. students and raise the question of whether students from countries other than the United States should be admitted to these programs. The judgment of MIT, even in these cases, is that as the programs mature, there should not be a prohibition on participation by international students, though it would be expected that U.S. students would naturally predominate.

These judgments about the nondiscrimination of foreign students once they are admitted are easier to make in part because a large fraction of foreign students, especially those from developing countries, elect to stay in the United States after graduation. This results in a net inflow that is of great importance to U.S. industry and universities and, correspondingly, often of concern to their home countries.

A survey of R&D directors of high-technology firms by the National Academy of Engineering, for example, showed that they believe their industries are dependent upon foreign talent, and that this dependency is increasing (NAE 1988: 3). Some 60 percent of foreign doctoral students intend to stay in the United States after graduation (NRC 1990: 46).[10] Engineering departments in U.S. universities are increasingly dependent on foreign students who stay on to accept faculty positions; in 1986, 47 percent of engineering faculty under 35 years old were not citizens or permanent residents (NAE 1988).

Moreover, those international students who do return home after graduation and go into industry tend to rise in their companies, universities, or government positions. The fact that future scientific, engineering, industrial and governmental leaders of other nations have spent their formative years in this country can provide a variety of benefits to the United States in the general advance of science, research collaboration, business relationships, and political interests. Those benefits can be important, but they will not be uniformly realized or predictable. It is instructive, however, to recall how important the nation believed was the payoff of the education of the future foreign elite during the Cold War.

The tendency of foreign students to stay in the United States has led to a resurgence of concern about a "brain drain" from developing countries. MIT makes essentially no provision for tailoring its engineering and science education to the needs of other countries, which adds to the likelihood that their students will be dissatisfied if they return home. In effect, MIT's education provides an incentive to stay in the United States. Other universities follow varied policies, but the general result is to encourage students to remain in the United States.

MIT believes that the unique quality of the MIT curriculum depends on its close interaction with the cutting edge of research. In fields where issues of particular concern to developing countries form important parts of the research agenda, such as urban studies, political science, and civil engineering, there is some focus on the Third World, with relevant courses and research opportunities. But in most fields at MIT, the domestic needs of developing countries are not directly addressed. This policy may be appropriate in light of the university's primary missions, but it can have significant negative effects on developing countries. However, the brain drain issue is not one that can be dealt with at the level of the university, but requires attention by the U.S. government and by the countries concerned.

Programs to Help U.S. Industry

It is an appropriate part of a U.S. university's mission to focus on problems of U.S. industry and on the effective transfer of knowledge to industry. Given the many challenging problems confronting the nation and the long-standing interest of many universities in effective transfer of knowledge, this responsibility need not be a diversion from the primary missions of education and research. The problems that arise from lackluster performance of U.S. industry in high-technology competition, and the danger of undesirable (and often self-defeating) policies that may result in reaction, mandate more explicit attention to this task than in the past.

Many programs that support this mission are now in place at the research universities; examples of a few that have emerged from faculty and administration initiatives at MIT in recent years illustrate promising directions. One is the MIT Commission on Industrial Productivity, which conducted detailed studies of eight major industries and offered a set of recommendations aimed at overcoming the slowdown in manufacturing productivity growth (Dertouzos et al. 1989). Another is the Leaders for Manufacturing program, conducted jointly by the Schools of Engineering and Management, which is intended to train a new generation of managers who understand production, a critical subject previously slighted by forefront universities in the United States. A third is the Microsystems Technology Laboratory, a $24 million facility for chip fabrication research supported by a consortium of U.S. firms.

A new Institute-wide research program will continue the study of productivity and industrial performance begun by the Productivity Commission. This new program will involve MIT faculty and students in a wide-ranging set of activities designed to bring the problems of U.S. industry into classrooms and laboratories, and will introduce ideas and discoveries generated at MIT into arenas of decision in industry and government.

An initiative of particular public policy interest—and of importance to the quality of MIT education as well—is the MIT/Japan Program. It is designed to give MIT students language and cultural skills and the experience of extended work in Japan, with the intention of helping to redress the imbalance in the flow of scientists and engineers and scientific and technological information between Japan and the United States. Some 250 students have completed the program since its inception in 1981. All students participating in the program learn Japanese, with additional courses preparing them for life in Japan. They then spend a year in a Japanese laboratory working as an engineer or scientist, or in a Japanese office as part of a management team. Forty-six MIT students were in residence in Japan during the 1990–1991 academic year—twenty-six in industrial laboratories, four in government facilities, and sixteen in universities. These students will be in a position throughout their careers to maintain relatively easy access to the scientific and technological community of Japan.

The MIT/Japan Program is a small start toward equalizing the technical exchange between the United States and Japan. But it is appreciable for a single university and it is also the largest program of its kind in the United States. Others have mounted their own efforts, and the MIT program formed the basis for a national initiative approved by Congress in 1990. MIT is considering building similar programs related to other nations and regions.

Programs centered on the needs of the U.S. economy can raise thorny problems of general principle at a university. First, how are these programs initiated and funded? Are they responses to faculty interest, or do university administrations attempt to start or impose them from above? The latter is an almost sure route to failure; yet, for an administration to be purely reactive is unlikely to produce any of the desired programs. Obviously, the answer is a subtle (and rare?) combination of incentives, resources, and intellectual leadership.

A second problem has been referred to earlier: whether foreign students should be allowed to take part in programs designed primarily to train U.S. students. The MIT solution is to allow an emphasis on U.S. students at the outset, but then to let nature take its course, without discrimination on the grounds of nationality.

A third problem is whether it is appropriate to limit the sponsorship of particular research efforts to U.S. companies. Occasionally, in developing projects at MIT intended to contribute to U.S. industrial objectives, consortia of companies have been formed to provide research funding that exclude participation of foreign companies. Such limited consortia provide important ways to fulfill national responsibilities and raise no issues of principle as long as: (1) faculty outside the program remain free to pursue their research with any sponsor, even if the subject is similar to, or in competition with, that of a consortium; (2) all the resulting research has similar conditions of access as other research at the Institute and will be freely and openly reported; and (3) there are no qualifications other than academic required of students who take a part in the research.

The issue of who is allowed access to research in progress in such U.S.-oriented programs becomes more visible as their special orientation to U.S. needs leads some to take the view that these research projects should be off limits to representatives of corporations from competitor nations. Presumably, however, at the major research universities it is assumed that all research, whatever its sponsorship, is conducted under the same rules of openness and access. Thus, foreign individuals should have the same ability to visit projects that address particular U.S. needs as any other project.

Some Key Questions

The experience of MIT and other research universities in fulfilling their multiple missions raises a number of concerns. This section summarizes issues that deserve the careful thought of government and university decision makers during this time of growing international technological competition.

1. *Asymmetry with Japan, especially in the ability to commercialize knowledge:* The question of asymmetry with Japan arises at every turn. It has many dimensions: the different relations and roles of industry and university research in Japan; the numbers of Japanese at U.S. universities and laboratories compared to U.S. citizens in Japan; differences in industrial structure and trade policies, which are the source of much friction between the countries; and many others. All of them make the university relationships to Japan a source of debate within the United States, and make reciprocity hard to achieve.

One particular asymmetry is politically most significant, and could be the proximate cause of restrictive legislation. It is the asymmetry in the ability of industry in the two countries to turn the results of research into commercial products. In the extreme, some assert that the game is lost—that whenever Japanese industry has equal access to research with U.S. companies, it will successfully dominate the resulting commercial market. With that view, a level playing field in terms of access to research turns automatically into a tilted field in the marketplace.

The implications of such a position are simply that Japanese companies must not be allowed equal access to research or that, if they are, there is no point in providing research support to U.S. universities since by so doing the nation is funding its competitors. This view was held by some who were well-placed in the U.S.

government in the 1980s. The improved performance of U.S. high-technology industry in the early 1990s, and the economic slowdown in Japan, may make this concern moot. But U.S. industry must still make major strides in its ability to commercialize knowledge for the issue to remain quiescent.

2. *Attention to critical fields:* Even if the general proposition above is rejected, are there any specific subjects with characteristics that would justify some modification of university practices of openness for a limited period of time? Such characteristics might be, for example, a particularly short jump from the laboratory to the marketplace, or "unique" importance in current economic competition, or some special meaning for proliferation of armaments. Applications of molecular biology might be such a field, or fundamental processes for manufacturing semiconductor chips, or subjects peculiarly tied to nuclear, biological, or chemical weapons.

3. *Extent of agreement on fundamental values of the research enterprise, and their actual implementation in practice:* The values and principles necessary for the performance of quality research are presented throughout this chapter as matters of firm agreement among academics, or at least as widely shared assumptions. Is there any dissent from these views? Are there situations or conditions under which some fraying of the principles would be both possible and acceptable? Do any examples of modified principles exist today at any of the research universities? The discussion above is from the perspective of MIT and similar universities. Do all universities that carry out substantial research follow the same principles?

It is also appropriate to ask whether, in practice, the general principles may be modified through implicit policies. Are qualified postdoctoral applicants chosen in all fields without any reference to nationality? Are funds for endowed chairs never raised with the likely incumbent previously identified? When an agreement is made to accept a long-term visitor from a company that has endowed a chair, how much actual control is exercised by the university in the selection of the visitor? Are all visitors to a research project treated in the same way in the transmission of knowledge, or is there de facto discrimination in the transfer of knowledge?

There is an important step between informal practices of some faculty, to the extent they exist, and the formal statement of rules and restrictions. If these modifications of strict interpretations of the value system of scientific research are at all common, what are the implications for a move by the government to attempt to impose formal restrictions?

4. *The possibility of foreign relationships unacceptable to universities on their own:* Are there any patterns of relationships with foreign governments and corporations that would be of sufficient concern to make universities willing to accept formal restrictions of some kind? For example, if foreign corporate support of research were not 1–2 percent of the total, or 3 percent as at MIT, but grew to be 30–50 percent, would that be a legitimate trigger for a change of policy? To take another example, at present U.S. universities would surely not accept research support in certain areas from some governments—support for nuclear research from Iraq is an obvious illustration. Would very large research support from Japan in economically sensitive subjects raise related concerns?

A related concern may be the growing number of Japanese corporate research laboratories being established in the United States near university centers. These well-financed labs will likely be magnets for the best U.S. scientists. It seems clear that in general, the expansion in the number of such laboratories in the United States, as long as they are operated as are comparable basic research labs of U.S. industry, is in the national interest by creating jobs in the United States, and keeping research at home. Would this be true if they grew in number and size so that they were siphoning substantial talent from U.S. universities?

These are just a few of the issues that attend the relations between research universities and foreign countries. This subject will undoubtedly be a source of much attention in the coming years, especially if high-technology trade relations with other nations, particularly Japan, do not improve. The likelihood that this issue will enter national politics as a subset of the larger trade question is high as well. Protectionist fervor remains strong despite the passage of the North American Free Trade Agreement (NAFTA) and the conclusion of the recent General Agreement on Tariffs and Trade (GATT) Uruguay round; moreover, it makes a good TV campaign issue, suitable for the requisite brief attack, while the response is necessarily lengthy and complex.

Universities need to be thoughtful and prepared. The issues involved are serious and legitimate. No adequate forum appears to exist for discussion among universities, nor between them and the government and industry. It is time one was created.

Acknowledgment

This chapter draws heavily on a study by a faculty committee of the Massachusetts Institute of Technology, of which the author was chairperson (Faculty Study Group 1991). A version of portions of this chapter appeared in "The American Research University," *Daedalus*, fall 1993, 122(4):225–252.

Notes

1. In some ways, international scientific and technological parity represents a return to the situation earlier in this century, when it was necessary for U.S. scientists to keep up with European developments in order to stay at the forefront of a field.

2. The casual use of "U.S." and "foreign" industry masks the blurring of the distinction between these terms. Their meaning is often in dispute: the definitions can depend on where the primary research and development centers are located, where the most value is added, the nationality of ownership, or where the headquarters is located. Nonetheless, the location of a firm's top management is still generally significant; and that is the definition that will be used in this chapter. For a discussion of the complexities of definition, see Reich (1990) and Tyson (1991).

3. The data and conclusions of a study of the international roles of research universities conducted at the Massachusetts Institute of Technology (MIT) form the basis for the discussion that follows (Faculty Study Group 1991). The issues are roughly similar for most research universities.

4. Some of the leading figures in U.S. science and technology policy have used the argument. See, for example, comments by former president of the National Academy of Sciences Frank Press (1990) and former director of the National Science Foundation Erich Bloch (1990: 12).

5. U.S. visitors are not tracked, except for official appointments.

6. It is worth noting that U.S. parochialism is a factor that inhibits learning about the work of others outside the United States. The "not-invented-here" syndrome can still bedevil the acquisition of knowledge from willing visitors, in the outdated belief that research in other countries could not be equivalent to that in the United States.

7. Though limited in volume, foreign research support is concentrated: five universities receive 51 percent of all foreign support; twenty receive 70 percent (NSB 1989).

8. Many faculty have observed that funding from foreign sources often comes with fewer strings and reporting requirements than does support from U.S. sources.

9. At least six chairs have been raised from Japanese industry for the Sloan School of Management on the argument that the students they send to that school have been implicitly subsidized in that tuition covers less than one-half the true cost of graduate education.

10. The available data on the rate at which international students stay in (and U.S. students leave) the United States are far from definitive, but they suggest that there has been no significant shift in the rate over the past decade.

Bibliography

Bank of Boston. 1989. *MIT: Growing Businesses for the Future.* Boston: Bank of Boston.

Bloch, E. R. 1990. University ties to foreign firms: letter. *Issues in Science and Technology* 6(4):12.

Chase Manhattan Corporation. 1990. *MIT Entrepreneurship in Silicon Valley.* New York: Chase Manhattan.

Dertouzos, M., Lester, R., and Solow, R. 1989. *Made in America: Regaining the Productive Edge.* Cambridge: MIT Press.

Faculty Study Group on the International Relationships of MIT. 1991. *The International Relationships of MIT in a Technologically Competitive World.* Cambridge: MIT.

General Accounting Office (GAO). 1988. *R&D Funding: Foreign Sponsorship of U.S. University Research.* GAO/RCED-88-89BR. Washington, DC: GAO.

National Academy of Engineering (NAE). 1988. *Foreign and Foreign-Born Engineers in the United States: Infusing Talent, Raising Issues.* Washington, DC: National Academy Press.

National Research Council (NRC). 1990. *Summary Report, 1989: Doctorate Recipients from United States Universities.* Washington, DC: National Academy Press.

National Science Board (NSB). 1989. *Report of the NSB Committee on Foreign Involvement in U.S. Universities.* Washington, DC: NSB.

National Science Foundation (NSF). 1990a. *National Patterns of R&D Resources: Funds and Personnel in the United States.* Washington, DC: NSF.

National Science Foundation (NSF). 1990b. *Selected Data on Academic Science and Engineering R&D Expenditures, Fiscal Year 1990.* Washington, DC: NSF.

National Science Foundation (NSF). 1993. *Foreign Participation in U.S. Academic Science and Engineering: 1991.* Special report. Washington, DC: NSF.

Press, F. 1990. Do the right thing. Address to members, annual meeting of the National Academy of Sciences, Washington, DC.

Reich, R. B. 1990. Who is us? *Harvard Business Review* 68(1):53–65.

Samuels, R., and Westney, D. E. 1987. Japanese scientific and technical information at MIT. Cambridge: MIT Center for International Studies.

Tyson, L. D. 1991. They are not us. *American Prospect* (4):37–49.

Westney, D. E. 1992. *Report on MIT Survey of Faculty International Relations.* Cambridge: MIT.

11

Conclusion: Constructive Responses to the Changing Social Context of University-Government Relations

David A. Hamburg

Introduction

In 1976, Sune Bergstrom, president of the Nobel Foundation, asked in his opening address at the Nobel Prize ceremonies why it was that Americans had swept all the prizes that year. Was it some kind of bicentennial conspiracy? He explained that the really interesting question was why Americans have been winning most of the Nobel Prizes for a quarter of a century, not why they won them all this particular year. He went on to elucidate what he perceived as some of the great strengths of U.S. science: the explicit, powerful recognition of basic science; the creation of great governmental institutions such as the National Institutes of Health and the National Science Foundation to foster basic science; the generous immigration policy that permitted the United States to absorb so many excellent German and other European and then Chinese scientists; and the decision to place most basic research in the universities. Young people received opportunities they did not get even in the democratic countries of Europe, even as late as 1976. He said that what Europeans should do is to emulate the U.S. experience, not complain about it.

What Bergstrom talked about was a rampant surge of scientific activity, including much institutional invention and institutional strengthening in this country during and after World War II. Before World War II, this country was not particularly distinguished in basic science. Scientists who wanted top-notch training in basic research traveled to Europe. Much of the new organization of science grew out of the Massachusetts Institute of Technology (MIT), Vannevar Bush being the great leader, but other MIT people being important as well. Especially after World War II, the U.S. consciousness had a vivid grasp of the unprecedented links between basic and applied research. There was the great (and in a way awful) recognition of the practical value of basic research through the atomic bomb. If you could build a Bomb, you could end a war. Those longhairs like Einstein must really have something to contribute!

I grew up in a small town in southern Indiana where, as across the nation, there was little interest in basic research. It was seen as long-haired stuff—arcane and impractical. But the Bomb made it very practical. Obviously other research was important, too, such as the work at MIT on radar. There were visible contributions by other universities: Columbia, Chicago, Berkeley, and others. These wartime technological tours-de-force made it much more attractive for the government to support basic science in a robust fashion, to place it in the universities, and to pay for the full costs, including substantial contributions to educating the next generation of scientists.

Basic science was perceived as an add-on to existing university responsibilities. The courses already existed for education—fundamentally undergraduate education, and for an elite sector of society at that. But the government was going to pay for it, and so had to broaden it in some unforeseeable ways. This was done for the good of the nation, not to be kind to MIT.

Science was not initially linked to the conflict with the Soviets. But a huge surge came, in my view, with the appraisal that the United States was going to require superior science and technology to win the Cold War. Of course, that was not the only motivation by any means. Certainly in the field of medicine, the enormous aspirations about improving public health fed strongly into the decision to support basic research. The discovery of antibiotics, soon followed by the polio vaccine, provided a potent inspiration. In any case, the system was in those days very small, and a few people around Vannevar Bush could make many of the key decisions. Science policy in the United States was run by a few committees, on very informal arrangements, and with a high degree of trust.

Changes and Criticism

What a difference a half-century makes! Then science was small, simple, and informal, based on trust. Now it is large, complex, and impersonal. There has been a diffuse loss of trust in science and technology and the universities, but also in much, much else. When I was president of the Institute of Medicine (IOM) from 1975 to 1980, I talked about the emerging fragility of the system. I thought it had grown so much and had become so complex, even then, that it was vulnerable to systemic errors and inflammatory rhetoric, like combustible material sitting around waiting for lightning to strike. I doubt if there is any institution that would not be damaged by hostile scrutiny of the sort that universities have borne in recent years. Given so many informal arrangements for the government-university relationship, based on trust in a small, simple system, errors cannot be avoided. These arrangements were highly susceptible to hostile attack and distortion. But certainly the system was mushy, particularly in the accounting and auditing functions; it was and remains to some degree vulnerable, although I think less so now than a few years ago. But this has not been the only focus of criticism.

What have been the major lines of attack? Research universities are said to be too elite, arrogant, expensive, unwilling to admit rank-and-file students, altogether a rich and privileged sector that generates envy and resentment in a lopsided society—a society that has become much more lopsided in the past dozen years than ever before in my lifetime. That lopsided character of the distribution of rewards and power in the society has predictably generated growing resentment. To some degree, the research universities, particularly the most elite ones, have suffered as a result. This has been a strong undertone of the congressional criticism.

Another criticism is related to the charge of arrogance. Professors are said to live an easy life and not to teach much. The large state universities are said to be ripping off the taxpayers. Legislators say in effect, "We put all the state money in, but the universities don't care whether the people they educate stay in the state. And anyway, they've got graduate assistants doing the teaching while they're in the laboratory."

Another line of attack is that the universities are not interested in local, area, or regional problems. Professors are interested in jetting around the world, while ignoring nearby suffering. Another criticism: universities are riddled with fraud and abuse and messy internal fights, even cover-ups. Another one: universities are "politically correct," dogmatic, rigid, ideological. And another: universities coddle some nasty and unqualified people—both faculty members and students. People who should not get in do get in; people who should not get promoted do get promoted. Then there's a stereotype that I remember from my youth: "These are eggheads doing impractical, arcane work that is remote from the lives of ordinary people."

Similar hostile criticisms of the scientific universities have been flowing for a long time, but mostly as undercurrents. There was a flood in the McCarthy period, and another flood in the "war on campus" period of the late 1960s and early 1970s. Why should there be a flood now?

The Social Context of the Current Criticism

During the 1980s, the nation became immersed in attitudes of selfishness, which took on a legitimacy and a pervasiveness which was extreme. The nation is not entirely out of it yet: a kind of here-and-now, no-tomorrow, devil-take-the-hindmost attitude is still widespread. Universities are fundamentally future-oriented, and I don't think they fare well in an atmosphere that is short-term and egocentrically oriented.

Pollsters and political consultants say they have never seen such pervasive public anger, at least not since the Depression. It is very hard to characterize. People seem to be disappointed in government, but they're also disappointed with almost all formerly respected institutions and almost all formerly respected individuals. Why should that be so severe right now? Whatever the underlying, predisposing factors may be, there is an exacerbation by the media. This negative orientation certainly got a boost from the betrayal of public trust and the deception by public officials during Vietnam and Watergate. One feels that the next Pulitzer Prize is just around the corner for ripping the mask off some putatively fraudulent university president or scientist, mayor or member of Congress.

But there is also a longer sweep of history. As far back as 1975, there were signs that the era of postwar domination by the United States was coming to an end; in a speech that year, I expressed concern that there might be far-reaching repercussions in society as the period of our preeminence gradually passed. To some extent, the current stress has its origins precisely in that remarkable transitional phase after World War II. With the image of a triumphant United States after the war, for a couple of decades there was a sense that science and technology would bring the nation great discoveries and innovations that would produce superb results. What kind of results? For one, military preeminence, even invincibility, at low cost. Second, economic prosperity, limitless horizons in time and money, and unending affluence unique in the history of the world. Third, improved health and drastically diminishing disease and disability. And finally, endless sexual pleasures, free of unwanted pregnancies, sexually transmitted diseases, or family concerns.

Of course, I am exaggerating. But I do believe this nation developed enormously high expectations that were doomed to serious, troubling disappointments. And in the face of an unexpected, distressing fall from grace, there is a pervasive human tendency to look for someone or something to blame. The research university is not the only scapegoat. There is a searching process to justify blame, because the achievements have not been as grand as anticipated. Despite the successful outcome of the Cold War, there is great disappointment abroad in the land in the realization that the multicentric world is far more complicated and frustrating than expected. Economic conditions have been only fair-to-middling. The cities are decaying. Sickness and disease remain. There is a groping for individuals and institutions to blame.

Constructive Responses

One of the great basketball clichés is, "Remember what got you here." If you are in the playoffs, build on the strengths that got you to the playoffs. The research universities of the United States are recognized throughout the world as outstanding institutions of truly global significance in education and research. What attributes made these research universities great? I suggest eight: (1) high standards of science and scholarship; (2) free and open inquiry; (3) objective methods of assessing information, ideas, and people; (4) respect for diversity in people and subject matter; (5) constant attention to opportunities for young people; (6) broad scope of coverage of subject matter on an in-depth basis; (7) a premium on the advancement of knowledge; and (8) a sense of social responsibility. These assets are all formidable, and universities should fight to retain them and build on them; these are the essential features that must be preserved.

But universities also need to do more. First, they must educate the public about their role in the nation's society and economy. Universities have tended to take for granted a special relationship with government and a special understanding on the part of the public; this may have been presumptuous. To the extent that universities can document that their vital functions are important for economic performance—for example, that research and development, and education and training stimulate economic performance—they need to make that widely understood. There is a mix of knowledge, skill, and freedom that is crucial for a vigorous democracy and for economic well-being. U.S. universities play a vital role in that mix.

Second, research universities should assume a kernel of truth in each accusation, until proven otherwise. Some charges are grotesque, but they can still have a small kernel. Charges must be taken seriously and examined reasonably. It is the only way universities can go forward with integrity—responding to the kernel of truth and the sober criticisms in a continuing way and, where necessary, making adjustments. Mistakes must be corrected. Systems that generate mistakes must be improved.

Third, university scholars should help the government sort out its organizations and its decision-making processes that deal with science and technology. That has been the focus of the Carnegie Commission on Science, Technology and Government. The Commission recommended new institutional arrangements that will put government in a better position to take science and technology into account in virtually all decision making in the next century. It has been an attempt to help the government get into a stronger position to deal with science and technology issues in an informed and deliberate way. The world has been transformed in the half-century since World War II, and the time is ripe for a reassessment of the government-university relationship.

Fourth, we must address the research university's role in life-long learning. With the world transformation of the economy, life-long learning is no longer a luxury but a requirement. Many people will have several careers, and continuous upgrading of technical competence is required in almost every field. The nation's population is aging, but people do not want to dry up and blow away; they want to keep learning and remain socially productive. They want to become students at all ages. Most universities will therefore need to develop vigorous outreach programs to involve a broader segment of the population directly in the life of the university. In the first instance, they can help to meet the national need for a technical economy in the world of the next century. Such educational outreach also builds the university constituency. It means many people, including influential people, would be directly engaged in the life of the university, not just for an occasional talk or fundraising pitch, but by virtue of upgrading their own knowledge and skills.

Finally, universities should have more relation to the "community"—local, national, and international. By "relation" I mean primarily interdisciplinary analytical work. The big problems of the world do not fit into the traditional packages of the disciplines, no matter how excellent the disciplines are. But analytical work should be rooted firmly in the disciplines, at the conjunction of the disciplines that tackle real-world problems in an analytical way, not a polemical way, with very high standards, including internal quality control mechanisms roughly akin to the important review mechanisms of the National Academy of Sciences (NAS).

A special art form of great value to policy makers and the world at large is the intelligible, credible synthesis. Elliot Richardson once told me that during his years in several cabinet positions he always wanted to know, "What are the facts?" The facts were scattered in many different places across a variety of fields, and were hard to sort out. He was exposed to clever briefers, advocacy groups, and special pleaders. How was he to form a solid, trustworthy, factual basis for policy decisions? In such situations, he badly needed a credible, intelligible synthesis—information he could understand as a layman, that he knew came from credible sources, that was vetted properly so that he could trust it. Universities are the obvious sources of such syntheses, which are needed today by many people besides cabinet officers. If universities want to meet this need, and thereby perform a valuable public service, they must place greater value on syntheses of information pertinent to real-world problems. At present, it may only be safe for tenured professors to work on such matters. But maybe universities could change the reward system so that brilliant young people could also spend some time on such problems.

Let us suppose that there is momentum gathering toward efforts of this kind, addressing great social problems, getting the facts straight, and performing objective analysis. What then? Some education beyond the campus could be very useful. For example, in arms control work at MIT and elsewhere, where major contributions have been made by university faculty, universities gradually came to realize that it was necessary to undertake education beyond the campus, meeting with government policy makers or publishing op-ed pieces or reports translated from technical into widely understandable language. After all, there are many socially useful sites for education. When universities reach beyond the campus, it is crucial to retain the high standards that exist on campus, and certainly not to be condescending.

So I believe that the research universities can make great contributions to the larger community over long time spans. On average, I think that such contributions would be appreciated, and would strengthen the position of universities in our society. I say this with the qualification that every important problem facing the society is, by definition, controversial. But there are ways to approach these issues that are not confrontational, that are respectful of different views, and that are analytical rather than polemical.

One special point should be made about involvement with the community. I believe that every university should have substantial links with the precollegiate schools in its geographic area. What form it should take I cannot prescribe. One program of the Carnegie Corporation, under the rubric "science rich/science poor," tries to link "science-rich" sectors of society such as the universities with "science-poor" sectors such as high schools, junior high schools, and elementary schools. In the long run, even the most elite universities will probably not get the desired quality of students unless they reach out to strengthen the primary and secondary schools in their areas. But beyond that objective is the possibility of stimulating a nationwide linkage system that could make a vital difference for the future of U.S. education. After all, there is a continuum of experience from prekindergarten to graduate education, and the whole range must be strong.

The research universities can also contribute, participate, even create fora to tackle some of the great problems that worry the U.S. public, including the relation of universities and government. As an example, think of the Government-University-Industry Research Roundtable at the NAS. There are university-based fora that work with, for example, the state congressional delegations. Universities like MIT, perhaps in cooperation with other universities in a national consortium, could undertake rolling reassessments of advances in science and technology and their implications for our economy and our society—conducting meetings across the country from time to time. Another example for possible future action would be a university-based follow-up to the Carnegie Commission on Science, Technology and Government.

The Carnegie Corporation has developed systematic linkages between scholars and the policy community. This effort links university-based researchers with U.S. policy makers and to some extent policy makers from other countries as well. A variety of formats are useful, the most in-depth involving a retreat setting of several days for which papers appropriate to the occasion are prepared and read by the participants in advance. Such meetings require balanced objectivity and low-key professionalism to be valuable. They can be fruitful if the topic is seen as crucial, the scientists and scholars are deeply informed, and the setting is conducive to the respectful exchange of information and ideas. University scientists and scholars are naturally major participants in such events. Linkage functions can be sponsored by universities, not only for policy makers but also for groups such as the media or the business community.

I believe that universities can usefully make linkages with the policy community and the media community at a very high level of quality—without arrogance and with a premium on getting the facts straight. Insofar as universities perform analytical work, it must be very fair-minded about the advantages and limitations of various policy-relevant options. For the research universities there is an important opportunity in fostering informed, deliberate, systematic communication with leaders in government and those who can educate the public accurately about the fascinating lines of inquiry and innovation that constitute the main thrusts of university life.

By the same token, the nature of the university itself—its structure and functions and its vital attributes—can be grist for the intellectual mill and for public understanding. In the end, a deep public trust is involved. Those of us who believe in the profound value of research universities must incessantly strive to be worthy of that trust.

Index